Passage to a Ringed World

The Cassini–Huygens Mission to Saturn and Titan

On the Cover: A collage of images shows the destination of the Cassini–Huygens mission — the Saturn system. The insets, from left, are Enceladus, Saturn and Titan. The collage represents the mission's five chief areas of scientific investigation: icy satellites, Saturn, Titan, rings and the magnetosphere. (Magnetospheric and plasma processes produce the "spokes" that run out across the rings as diffuse, dark markings.)

Passage to a Ringed World

The Cassini–Huygens Mission to Saturn and Titan

Linda J. Spilker, *Editor*

Jet Propulsion Laboratory

California Institute of Technology

NASA SP-533

National Aeronautics and Space Administration

Washington, D.C.

October 1997

This book is for you. You will find that it lives up to the promise of its cover and title. The text is authoritative, but at the same time easy to understand. It is written for any layperson who is interested in space exploration. In this book, we will give you a good look at what is involved in sending a large spacecraft to the outer solar system. Join us and share in the excitement of this extended voyage of discovery!

One of the first things you will learn is that Cassini–Huygens is an international mission. Seventeen countries are involved. You will also learn that Huygens is an atmospheric probe that the Cassini spacecraft will deliver to Titan, the largest moon of Saturn. Titan has a gaseous atmosphere that is thicker and more obscuring than the atmosphere here on Earth. We will look through Titan's atmosphere with Cassini's radar and discover surface details for the first time. The mission will find out if there are really liquid hydrocarbons on Titan's surface in the form of lakes or seas.

The Cassini spacecraft itself will spend four years in orbit about Saturn. We will examine the rings and visit many of the satellites. We will sample special locations in the magnetosphere that are believed to harbor interesting – some might say strange – plasma processes (that is, electromagnetic interactions involving electrons, protons and ions).

You can see in the table of contents the "menu" we have prepared for you. One chapter explains the mission, another the spacecraft. Other chapters tell you about Saturn, Titan, the rings and the various other parts of the Saturn system. These chapters reflect the facts and theories as we know them today. The chapters were prepared by expert researchers in each of the areas covered. Professional writers then edited the text for clarity and to make sure that you would not be overwhelmed with jargon and technospeak. Working hand-in-hand with them were graphic designers and illustrators who created the

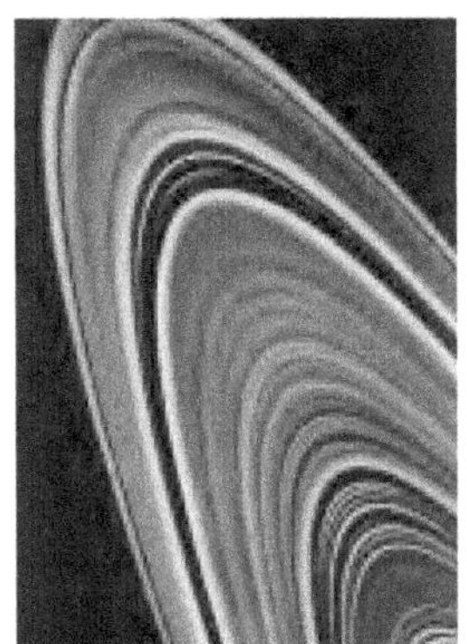

page layouts and the many informative diagrams. Finally, to assure that the final product was accurate, the book was reviewed by scientists and engineers who work on the Cassini–Huygens mission.

Come with us now as we set out on this voyage of exploration. We have over three billion kilometers to go! It will take more than six years to reach Saturn. We will get there on July 1, 2004. By the end of the Cassini–Huygens mission, four years later, we will have completed the most complicated scientific experiment ever performed.

We wish you good reading!

Richard J. Spehalski Cassini Program Manager	Dennis L. Matson Cassini Project Scientist

ACKNOWLEDGMENTS

This book would not have been possible without the outstanding efforts of many talented people, from the authors who wrote the chapters, to the editors and reviewers, to the graphic designers, artists and finally the printer. I would like to thank all of them for their excellent work. Many long hours went into developing the exciting Cassini–Huygens book you hold in your hand. Cassini–Huygens plans to reap a bountiful harvest of new information about the Saturn system, so stay tuned!

Many individuals worked diligently on writing the chapters in this book. These dedicated authors include Vince Anicich, Scott Bolton, Jim Bradley, Bonnie Buratti, Marcia Burton, Roger Diehl, Stephen Edberg, Candy Hansen, Charley Kohlhase, Paulett Liewer, Dennis Matson, Ellis Miner, Nicole Rappaport, Laci Roth, Linda Spilker and Brad Wallis. I also thank Stephen Edberg, Charley Kohlhase, Ellis Miner and Randii Wessen for their support in creating the Appendices.

The creative design of this book is the ingenious work of the Jet Propulsion Laboratory's Design Services team. The team worked tirelessly to complete this work before the Cassini–Huygens launch. My thanks to Audrey Steffan-Riethle for design and art direction. Special thanks go to Sanjoy Moorthy who spent many long hours writing and editing and organizing the material. My thanks also to David Hinkle for illustration and production, David Seal for computer artwork, Adriane Jach and Elsa King-Ko for design and production, Marilyn Morgan for editing and Robert Chandler for printing coordination.

Many people provided reviews of the chapters. Special thanks for detailed reviews and editorial support go to Stephen Edberg, Charley Kohlhase, Jean-Pierre Lebreton, Ellis Miner, Nicole Rappaport, Laci Roth and Randii Wessen. My thanks to Marjorie Anicich, Michel Blanc, Jeff Cuzzi, Luke Dones, Michele Dougherty, Charles Elachi, Larry Esposito, Bill Fawcett, Ray Goldstein, Don Gurnett, Len Jaffe, Bill Kurth, Janet Luhman, Jonathan Lunine, Carl Murray, Andy Nagy, Toby Owen, Carolyn Porco, Dave Young and Philippe Zarka for serving as reviewers.

Linda J. Spilker

Linda J. Spilker
Cassini Deputy Project Scientist

Contents

CHAPTER 1

A Mission of Discovery

Saturn, the second most massive planet in the solar system, offers us a treasure of opportunities for exploration and discovery. Its remarkable system of rings is a subject intensively studied, described and cataloged. Its planet-sized satellite, Titan, has a dense, veiled atmosphere. Some 17 additional icy satellites are known to exist — each a separate world to explore in itself. Saturn's magnetosphere is extensive and maintains dynamic interfaces with both the solar wind and with Titan's atmosphere.

From Observation to Exploration

Saturn was known even to the ancients. Noting the apparitions of the planets and recording other celestial events was a custom of virtually every early civilization. These activities knew no oceanic bounds, suggesting a curiosity universal to humankind. Real progress in understanding the planets came with the invention and use of instruments to accurately measure celestial positions, enabling astronomers to catalog the planets' positions.

Developments in mathematics and the discovery of the theory of gravitational attraction were also necessary steps. With the invention of the telescope and its first application in observing the heavens by Galileo Galilei, the pace of progress quickened.

Three separate images, taken by Voyager 2 through ultraviolet, violet and green filters, respectively, were combined to create this false-color image of Saturn.

Today, the Cassini–Huygens mission to Saturn and Titan is designed to carry out in-depth exploration of the Saturn system. A payload of science instruments will make in situ measurements or observe their targets under favorable geometric and temporal circumstance. For example, an instrument might make observations over a range of various angles of illumination and emission, or study events such as occultations or eclipses.

Cassini–Huygens' interplanetary journey starts in October 1997, with the launch from Cape Canaveral in Florida. Upon arrival at Saturn, Cassini–Huygens will go into orbit about the planet. The spacecraft consists of two parts — the Cassini Orbiter and a smaller spacecraft, the Huygens Probe, which is targeted for Titan, Saturn's largest moon.

Huygens will arrive at Titan in November 2004. After using its heat shield for deceleration in Titan's upper atmosphere, Huygens will deploy a parachute system. Six instruments will make scientific measurements and observations during the long

GETTING TO KNOW SATURN
A TIMELINE OF DISCOVERY

~ 800 BC	Assyrian and Babylonian observations
~ 300 AD	Mythological view of Saturn the god
1610	Galileo notes the "triple planet" Saturn with his telescope
1655–59	Huygens discovers Saturn's largest satellite, Titan
1671–84	Cassini discovers a division in the ring; he also discovers the satellites Iapetus, Rhea, Dione and Tethys
1789	Herschel discovers satellites Mimas and Enceladus – and notes thinness of rings
1848	Bond and Lassel discover the satellite Hyperion
1850	Bond, Bond and Daws discover inner ring
1857	Maxwell proves that rings are not solid
1895	Keeler measures ring velocities
1898	Pickering discovers satellite Phoebe
1932	Wildt discovers methane and ammonia on Saturn
1943–44	Kuiper discovers methane and ammonia on Titan
1979	Pioneer 11 flies past Saturn
1980	Voyager 1 encounters Saturn
1981	Voyager 2 encounters Saturn
1989	Hubble Space Telescope's Wide Field and Planetary Camera images Saturn
1995	Wide Field and Planetary Camera 2 images ring plane crossing
1997	Cassini–Huygens launches
2004	Cassini–Huygens enters Saturn orbit; Huygens explores Titan

descent to the surface. The Huygens Probe data will be transmitted to the Orbiter and then to Earth.

The Orbiter then commences a tour of the Saturn system. With its complement of 12 instruments, Cassini is capable of making a wide range of in situ and remote-sensing observations. The Orbiter will make repeated close flybys of Titan to make measurements and obtain observations.

The flybys of Titan will also provide gravity-assisted orbit changes, enabling the Orbiter to visit other satellites and parts of the magnetosphere and observe occultations of the rings and atmospheres of Saturn and Titan. Over the span of the four-year-long orbital mission, Cassini is expected to record temporal changes in many of the properties that it will observe.

Vision for a Mission

The Cassini–Huygens mission honors two astronomers who pioneered modern observations of Saturn. The Orbiter is named for Jean-Dominique Cassini, who discovered the satellites Iapetus, Rhea, Dione and Tethys, as well as ring features such as the Cassini division, in the period 1671–1684. The Titan Probe is named for Christiaan Huygens, who discovered Saturn's largest satellite in 1655.

The mission is a joint undertaking by the National Aeronautics and Space Administration (NASA) and the European Space Agency (ESA). The Huygens Probe is supplied by ESA and the main spacecraft — the Orbiter — is provided by NASA. The Italian space agency (Agenzia Spaziale Italiana, or ASI), through a bilateral agreement with NASA, is providing hardware systems for the Orbiter spacecraft and instruments. Other instruments on the Orbiter and the Probe are provided by scientific groups and/or their industrial partners, supported by NASA or by the national funding agencies of member states of ESA. The launch vehicle and launch operations are provided by NASA. NASA will also provide the mission operations and telecommunications via the Deep Space Network (DSN). Huygens operations are carried out by ESA from its operations center in Darmstadt, Germany.

Late in 1990, NASA and ESA simultaneously selected the payloads for the Orbiter and for the Huygens Probe, respectively. Both agencies also selected interdisciplinary investigations. The NASA Orbiter selection comprises seven principal investigator instruments, five facility instruments and seven interdisciplinary investigations. The ESA Huygens selection comprises six principal investigator instruments and three interdisciplinary scientist investigations.

This complex, cooperative undertaking did not come into being overnight. Rather, it was the end product of a process of joint discussions and careful planning. The result — the Cassini–Huygens mission — is an enterprise that, from the initial vision to the completion of the nominal mission, will span nearly 30 years! The formal beginning was in 1982, when a Joint Working Group was formed by the Space Science Committee of the European Science Foundation and the Space Science Board of the National Academy of Sciences in the United States.

The charter of the group was to study possible modes of cooperation between the United States and Europe in the field of planetary science. The partners were cautious and did not enter lightly into the decision to carry out the Cassini–Huygens mission. Their precept was that the mission would be beneficial for the scientific, technological and industrial sectors of their countries.

The Quest for Understanding

In carrying out this voyage, we are following a basic, evolutionally nurtured instinct to explore our environment. Whether exploration results in the discovery of resources or the recognition of hazards, or merely provides a sense of place or accomplishment, it has always proved beneficial to be familiar with our environment.

It is not surprising, therefore, that such exploration is a hallmark of growing, thriving societies. Parallels can be drawn between historical voyages of exploration and the era of solar system exploration. Available technology, skilled labor, possible benefits, cost, risk and trip duration continue to be some of the major considerations in deciding — to go or not to go? All these factors were weighed for Cassini–Huygens — and we decided to go.

In traveling to Saturn with Cassini–Huygens, we will also be satisfying a cultural desire to obtain new knowledge. This drive is very strong because modern society places a very high value on knowledge. For example, the whole field of education is focused on transferring knowledge to new generations. Often, a substantial portion of a person's existence is spent in school. The resources expended in creating new knowledge through inventions, research and scholarship are considered to be investments, with the beneficial return to come later. Knowledge is advantageous to us.

With Cassini–Huygens, we do not have to wait for our arrival at Saturn, because the return of new knowledge has already occurred. The challenge of this mission has resulted in new technological developments and inventions. Some of these have already been spun off to new applications, and new benefits are being realized now. But as with any investment, the main return is expected later, when Cassini–Huygens carries out its mission at Saturn.

Cassini–Huygens is the next logical step in the exploration of the outer solar system. The Jupiter system has been explored by Galileo. Now it is Saturn's turn. With Cassini–Huygens, we will explore in depth a new part of the solar system. Not only will we learn about the Saturn system, but as a result we will also learn more about Earth as a part of the solar system — rather than as an isolated planet.

Physics and chemistry are the same everywhere. Thus, knowledge gained about Saturn's magnetosphere or Titan's atmosphere will have application here on Earth. Interactions with the solar wind and impacts of comets and asteroids are just two of the processes that planets have in common. Information gleaned at Saturn about these shared histories and processes will also lead us to new information about Earth and its history.

The Cassini–Huygens mission also allows the realization of other goals. Bringing people of different countries together to work toward a common goal promotes understanding and common values. The international character of Cassini–Huygens permits talented engineers and scientists to

VEHICLE OF DISCOVERY

After a deep space voyage, Cassini will spend years studying the vast Saturn system.

A big mission requires a big spacecraft — and Cassini fits the bill. When Orbiter and Probe are attached and fueled, the spacecraft weighs over 5600 kilograms.

Its body stands almost seven meters tall and is over four meters wide. Here the spacecraft is shown without its thermal blankets.

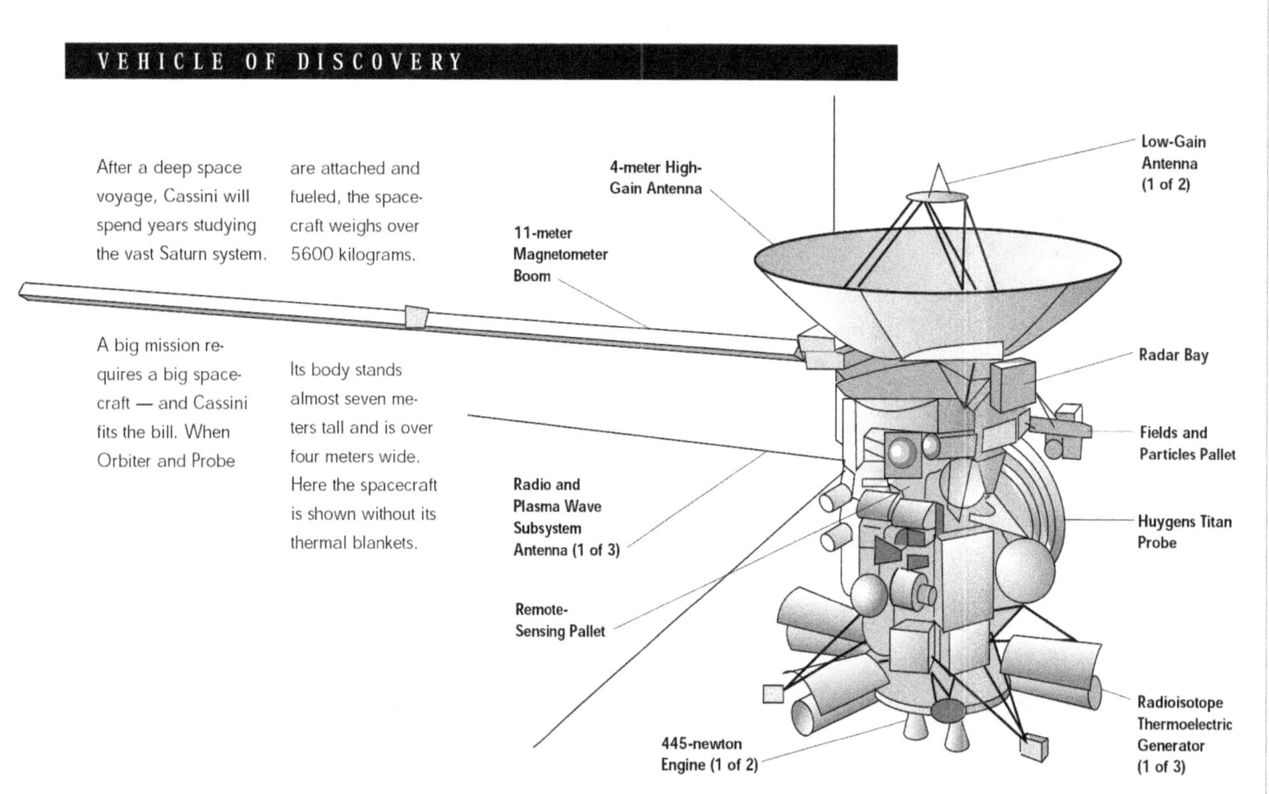

address the challenges of the mission. More people will share in the discoveries, the costs and the benefits.

About three quarters of a million people from more than 80 countries have involved themselves in the Cassini–Huygens mission by requesting that their signatures be placed aboard the spacecraft for the trip to the Saturn system. What will the mission bring to these people? The answers are varied: satisfaction of curiosity, a sense of participation, aesthetic inspiration, perhaps even understanding of our place in this vast universe.

The Spacecraft

At the time of launch, the mass of the fully fueled Cassini spacecraft will be about 5630 kilograms. Cassini consists of several sections. Starting at the bottom of the "stack" and moving upward, these are the lower equipment module, the propellant tanks together with the engines, the upper equipment module, the 12-bay electronics compartment and the high-gain antenna. These are all stacked vertically on top of each other. Attached to the side of the stack is an approximately three-meter-diameter, disk-shaped spacecraft — the Huygens Titan Probe. Cassini–Huygens accommodates some 27 scientific investigations, supported by 18 specially designed instruments: 12 on the Orbiter and six on the Probe.

Most of the Orbiter's scientific instruments are installed on one of two body-fixed platforms — the remote-sensing pallet or the fields and particles pallet — named after the type of instruments they support. The big, 11-meter-long boom supports sensors for the magnetometer experiment. Three thin 10-meter-long electrical antennas point in orthogonal directions; these are sensors for the Radio and Plasma Wave Science experiment. At the top of the stack is the large, four-meter-diameter high-gain antenna. Centered and at the very top of this antenna is a relatively small low-gain antenna. A second low-gain antenna is located near the bottom of the spacecraft.

Two-way communication with Cassini will be through NASA's Deep Space Network (DSN) via an X-band radio link, which uses either the four-meter-diameter high-gain antenna or one of the two low-gain antennas. The high-gain antenna is also used for radio and radar experiments and for receiving signals from Huygens.

The electrical power for the spacecraft is supplied by three radioisotope thermoelectric generators. Cassini is a three-axis-stabilized spacecraft. The attitude of the spacecraft is changed by using either reaction wheels or the set of 0.5-newton thrusters. Attitude changes will be done frequently because the instruments are body-fixed and the whole spacecraft must be turned in order to point them. Consequently, most of the observations will be made without a real-time communications link to Earth. The data will be stored on two solid-state recorders, each with a capacity of about two gigabits.

Scientific data will be obtained primarily by using one or the other of two modes of operation. These modes have been named after the functions they will carry out and are called the "remote-sensing mode" and the "fields and particles and downlink mode."

During remote-sensing operations, the recorders are filled with images and spectroscopic and other data that are obtained as the spacecraft points to various targets. During the fields and particles and downlink mode, the high-gain antenna is pointed at Earth and the stored data are transmitted to the DSN. Also, while in this mode, the spacecraft is slowly rolled about the axis of the high-gain antenna. This allows sensors on the fields and particles pallet to scan the sky and determine directional components for the various quantities they measure.

Mission Overview

The Cassini–Huygens mission is designed to explore the Saturn system and all its elements — the planet Saturn and its atmosphere, its rings, its magnetosphere, Titan and many of the icy satellites. The mission will pay special attention to Saturn's largest moon, Titan, the target for Huygens.

The Cassini Orbiter will make repeated close flybys of Titan, both for gathering data about Titan and for gravity-assisted orbit changes. These maneuvers will permit the achievement of a wide range of desirable characteristics on the individual orbits that make up the tour. In turn, this ability to change orbits will enable close flybys of icy satellites, re-

CASSINI-HUYGENS SCIENCE INVESTIGATIONS

	Instrument	Investigations
Cassini Saturn Orbiter Fields and Particles Instruments	Cassini Plasma Spectrometer	In situ study of plasma within and near Saturn's magnetic field
	Cosmic Dust Analyzer	In situ study of ice and dust grains in the Saturn system
	Dual Technique Magnetometer	Study of Saturn's magnetic field and interactions with the solar wind
	Ion and Neutral Mass Spectrometer	In situ study of compositions of neutral and charged particles within the magnetosphere
	Magnetospheric Imaging Instrument	Global magnetospheric imaging and in situ measurements of Saturn's magnetosphere and solar wind interactions
	Radio and Plasma Wave Science	Measurement of electric and magnetic fields, and electron density and temperature in the interplanetary medium and within Saturn's magnetosphere
Cassini Saturn Orbiter Remote-Sensing Instruments	Cassini Radar	Radar imaging, altimetry, and passive radiometry of Titan's surface
	Composite Infrared Spectrometer	Infrared studies of temperature and composition of surfaces, atmospheres and rings within the Saturn system
	Imaging Science Subsystem	Multispectral imaging of Saturn, Titan, rings and icy satellites to observe their properties
	Radio Science Instrument	Study of atmospheric and ring structure, gravity fields and gravitational waves
	Ultraviolet Imaging Spectrograph	Ultraviolet spectra and low-resolution imaging of atmospheres and rings for structure, chemistry and composition
	Visible and Infrared Mapping Spectrometer	Visible and infrared spectral mapping to study composition and structure of surfaces, atmospheres and rings
Huygens Titan Probe Instruments	Aerosol Collector and Pyrolyser	In situ study of clouds and aerosols in Titan's atmosphere
	Descent Imager and Spectral Radiometer	Measurement of temperatures of Titan's atmospheric aerosols and surface imagery
	Doppler Wind Experiment	Study of winds by their effect on the Probe during descent
	Gas Chromatograph and Mass Spectrometer	In situ measurement of chemical composition of gases and aerosols in Titan's atmosphere
	Huygens Atmospheric Structure Instrument	In situ study of Titan's atmospheric physical and electrical properties
	Surface Science Package	Measurement of the physical properties of Titan's surface

Participant Nations*

* *First flag in each row represents nation of Principal Investigator or Team Leader.*

connaissance of the magnetosphere over a variety of locations and the observation of the rings and Saturn at various illumination and occultation geometries and phase angles.

The spacecraft will be injected into a 6.7-year Venus–Venus–Earth–Jupiter Gravity Assist (VVEJGA) trajectory to Saturn. Included are gravity assists from Venus (April 1998 and June 1999), Earth (August 1999) and Jupiter (December 2000). Arrival at Saturn is planned for July 2004.

During most of the early portion of the cruise, communication with the spacecraft will be via one of the two low-gain antennas. Six months after the Earth flyby, the spacecraft will turn to point its high-gain antenna at Earth and communications from then on will use the high-gain antenna.

Following the Jupiter flyby, the spacecraft will attempt to detect gravitational waves using Ka-band and X-band radio equipment. Instrument calibrations will also be done during cruise between Jupiter and Saturn. Science observations will begin two years away from Saturn (about one and a half years after the Jupiter flyby).

The most critical phase of the mission following launch is the Saturn orbit insertion (SOI) phase. Not only will it be a crucial maneuver, but it will also be a period of unique scientific activity, because at that time the spacecraft will be the closest it will ever be to the planet.

The SOI phase of the trajectory will also provide a unique opportunity for observing the rings. The spacecraft's first orbit will be the longest in the orbital tour. A periapsis-raise maneuver in September 2004 will establish the geometry for the Huygens Probe entry at the spacecraft's first Titan flyby in November.

Huygens' Encounter with Titan. In November 2004, the Huygens Probe will be released from the Cassini Orbiter, 21–22 days before the first Titan flyby. Two days after the Probe's release, the Orbiter will perform a deflection maneuver; this will keep the Orbiter from following Huygens into Titan's atmosphere. It will also establish the required radio-communication geometry between the Probe and the Orbiter, which is needed during the Probe descent phase, and will also set the initial conditions for the satellite tour — which starts right after the completion of the Probe mission.

The Huygens Probe has the task of entering Titan's atmosphere, making in situ measurements of the satellite's properties during descent by parachute to the surface. The Probe consists of a descent module enclosed by a thermal-protection shell. The front shield of this shell is 2.7 meters in di-

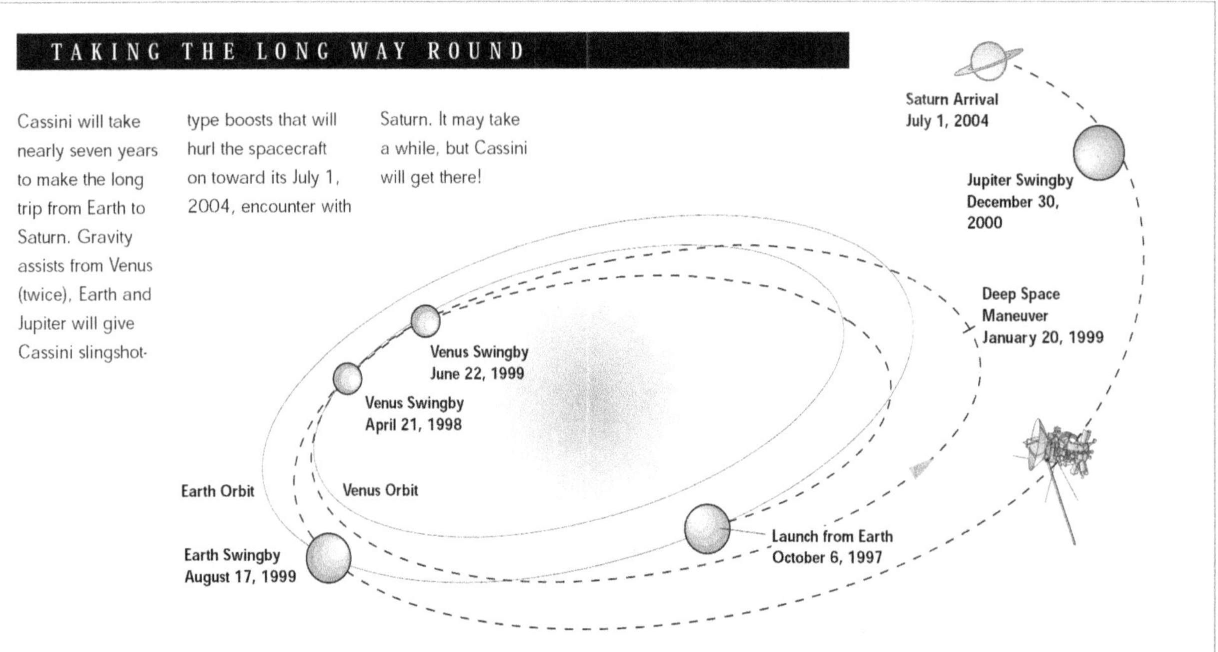

PROBING A MOON OF MYSTERY

Huygens' encounter with Titan is planned for November 2004, about three weeks before the Orbiter makes its first flyby of Saturn's largest satellite. The Probe, which consists of a descent module enveloped by a conical thermal-protection shell almost three meters in diameter, carries six science instruments designed to make in situ measurements of Titan's atmospheric and surface properties.

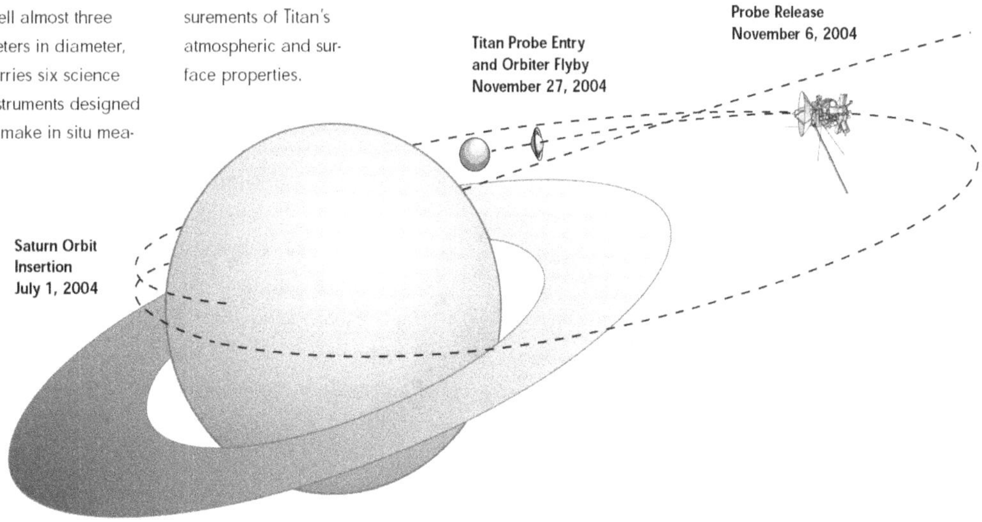

ameter and is a very bluntly shaped conical capsule with a high drag coefficient. The shield is covered with a special thermal-ablation material to protect the Probe from the enormous flux of heat generated during atmospheric entry. On the aft side is a protective cover that is primarily designed to reflect away the heat radiated from the hot wake of the Probe as it decelerates in Titan's upper atmosphere. Atmospheric entry is a tricky affair — entry at too shallow an angle can cause the Probe to skip out of the atmosphere and be lost. If the entry is too steep, that will cause the Probe measurements to begin at a lower altitude than is desired.

After the Probe has separated from the Orbiter, the electrical power for its whole mission is provided by five lithium–sulfur dioxide batteries. The Probe carries two S-band transmitters and two antennas, both of which will transmit to the Orbiter during the Probe's descent. One stream of telemetry is delayed by about six seconds with respect to the other to avoid data loss if there are brief transmission outages.

Once the Probe has decelerated to about Mach 1.5, the aft cover is pulled off by a pilot parachute. An 8.3-meter-diameter main parachute is then deployed to ensure a slow and stable descent. The main parachute slows the Probe and allows the decelerator and heat shield to fall away.

To limit the duration of the descent to a maximum of two and a half hours, the main parachute is jettisoned at entry +900 seconds and replaced by a smaller, three-meter-diameter drogue chute for the remainder of the descent. The batteries and other resources are sized for a maximum mission duration of 153 minutes. This corresponds to a maximum descent time of two and a half hours, with at least three minutes — but possibly up to half an hour or more — on the surface, if the descent takes less time than expected.

The instrument operations are commanded by a timer in the top part of the descent and on the basis of measured altitude in the bottom part of the descent. The altitude is measured by a small radar altimeter during the last 10–20 kilometers.

Throughout the descent, the Huygens Atmospheric Structure Instrument (HASI) will measure more than half a dozen physical properties of the atmosphere. The HASI will also process signals from the Probe's radar altimeter to gain information about surface properties. The Gas Chromatograph and Mass Spectrometer (GCMS) will determine the chemical composition of the atmosphere as a function of altitude. The Aerosol Collector and Pyrolyser (ACP) will capture aerosol particles, heat them and send the effused gas to the GCMS for analysis.

The optical radiation propagation in the atmosphere will be measured in all directions by the Descent Imager and Spectral Radiometer (DISR). The DISR will also image the cloud formations and the surface. As the surface looms closer, the DISR will switch on a bright lamp and measure the spectral reflectance of the surface.

Throughout its descent, the Doppler shift of Huygens' telemetric signal will be measured by the Doppler Wind Experiment (DWE) equipment on the Orbiter to determine the atmospheric winds, gusts and turbulence.

In the proximity of the surface, the Surface Science Package (SSP) will activate a number of its devices to make measurements near and on the surface. If touchdown occurs in a liquid, such as in a lake or a sea, the SSP will measure the liquid's physical properties.

The Orbital Tour. After the end of the Probe mission, the Orbiter will start its nearly four-year tour, consisting of more than 70 Saturn-centered orbits, connected by Titan-gravity-assist flybys or propulsive maneuvers. The size of these orbits, their orientation to the Sun–Saturn line and their inclination to Saturn's equator are dictated by the various scientific requirements, which include Titan ground-track coverage; flybys of icy satellites, Saturn, Titan, or ring occultations; orbit inclinations; and ring-plane crossings.

DESCENT INTO THE MURKY DEPTHS

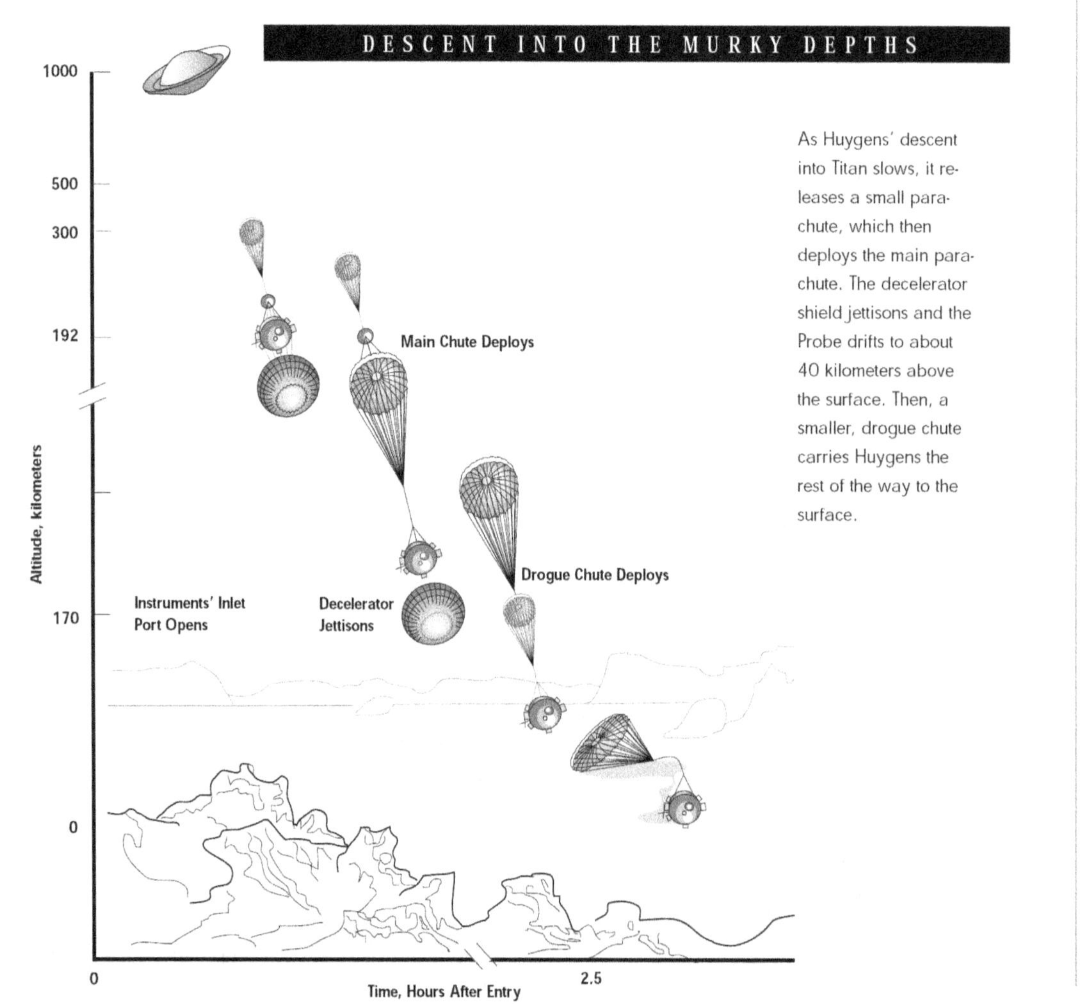

As Huygens' descent into Titan slows, it releases a small parachute, which then deploys the main parachute. The decelerator shield jettisons and the Probe drifts to about 40 kilometers above the surface. Then, a smaller, drogue chute carries Huygens the rest of the way to the surface.

TOUR T18-3
TARGETED SATELLITE FLYBYS*

Cassini Flybys, Planned	Saturn Satellite	Voyager Closest Approach, kilometers
1	Iapetus	909,000
4	Enceladus	87,000
1	Dione	162,000
1	Rhea	74,000

** Actual distances will be based on the type of observations planned.*

TOUR T18-3
SERENDIPITOUS SATELLITE FLYBYS*

Saturn Satellite	5000–25,000 kilometers	25,000–50,000 kilometers	50,000–100,000 kilometers	Voyager Flyby Range, kilometers
Mimas	–	1	4	88,000
Enceladus	3	–	3	87,000
Tethys	1	3	6	93,000
Dione	–	2	2	161,00
Rhea	–	1	1	74,000
Phoebe	–	1	–	2,076,000

** For flybys with closest approach on the sunlit side.*

Titan is the only Saturn satellite that is large enough to enable significant gravity-assisted orbit changes. The smaller icy satellites can help sometimes with their small perturbations, which can be useful in trimming a trajectory. Designing the Cassini orbital tour is a complicated and challenging task that will not be completed for at least several years.

One of the tours under consideration (called T18-3) can be used to illustrate the complexity involved in this type of navigational planning. T18-3 was the eighteenth tour designed for Cassini: This example is the third iteration of that tour. Tour T18-3 comprises 43 Titan flybys and seven "targeted" flybys of the icy satellites Iapetus, Enceladus, Dione and Rhea. "Targeted" means that the flyby distance can be chosen to best accommodate the planned observations — generally in the 1000-kilometer range. The closest approaches to the satellites made by the Voyager spacecraft in the early 1980s are used for reference.

In addition to the targeted flybys, there are other, unplanned — serendipitous — flybys that occur as a result of the tour path's geometry. If these flybys are close enough to the satellites, they will provide valuable opportunities for scientific observations. In our sample tour, T18-3, there are 28 of these flybys, with distances of less than 100,000 kilometers. At a range of 100,000 kilometers, the pixel resolution of Cassini's narrow-angle camera is about 0.6 kilometer for several of the icy satellites, providing better resolution than that achieved by Voyager.

Science Objectives

All the activities on Cassini–Huygens are pointed toward the achievement of the mission's science objectives. The origin of the objectives can be traced all the way back to the meetings of the Joint Working Group in 1982. They were further developed during the Joint NASA–ESA assessment study in 1984-85. As a result of this study, the science objectives appeared for the first time in their present form in the group's final report, issued by ESA in 1985. The objectives then became formally established with their inclusion in the NASA and ESA announcements of opportunity (ESA, 1989; NASA, 1989, 1991). The science objectives for Cassini–Huygens are organized by mission phase and target.

Scientific Objectives During Cruise. Given the trajectory for the long voyage to Saturn, Cassini–Huygens has the opportunity to carry out a number of experiments during the cruise phase. After launch, there will be instrument checkouts and maintenance activities. Searches for gravity waves will be carried out during the three successive oppositions of the spacecraft, beginning in December 2001.

These searches are radio experiments that involve using the DSN for two-way, K_a–band tracking of the spacecraft. During two solar conjunctions of the spacecraft, a series of radio-propagation measurements obtained from two-way X-band and K_a-band DSN tracking will provide a test of general relativity, as well as data on the solar corona.

In early 1992, the Cassini project team at the Jet Propulsion Laboratory redesigned the mission and the Orbiter to meet NASA budgetary constraints expected for the following years. As a result, most of the science objectives for targets of opportunity — asteroid flyby, Jupiter flyby and cruise — were deleted. This was part of the effort to control development costs and the cost of operation during the first few years in flight. In the current baseline plan, the scientific data acquisition will start two years before arrival at Saturn; that is, well after the Jupiter flyby.

Scientific Objectives at Saturn. The list of scientific objectives for Cassini–Huygens is extensive. There are specific objectives for each of the types of bodies in the system — the planet itself, the rings, Titan, icy satellites and the magnetosphere. Not only is

ENCOUNTERS WITH A CELESTIAL GIANT

The actual makeup of Cassini's four-year "tour" of the Saturn system is yet to be finalized, but many possible scenarios are already under examination.

This sample tour, named T18-3, contains over 70 orbits of Saturn, over 40 flybys of Titan and a number of close flybys of several other satellites.

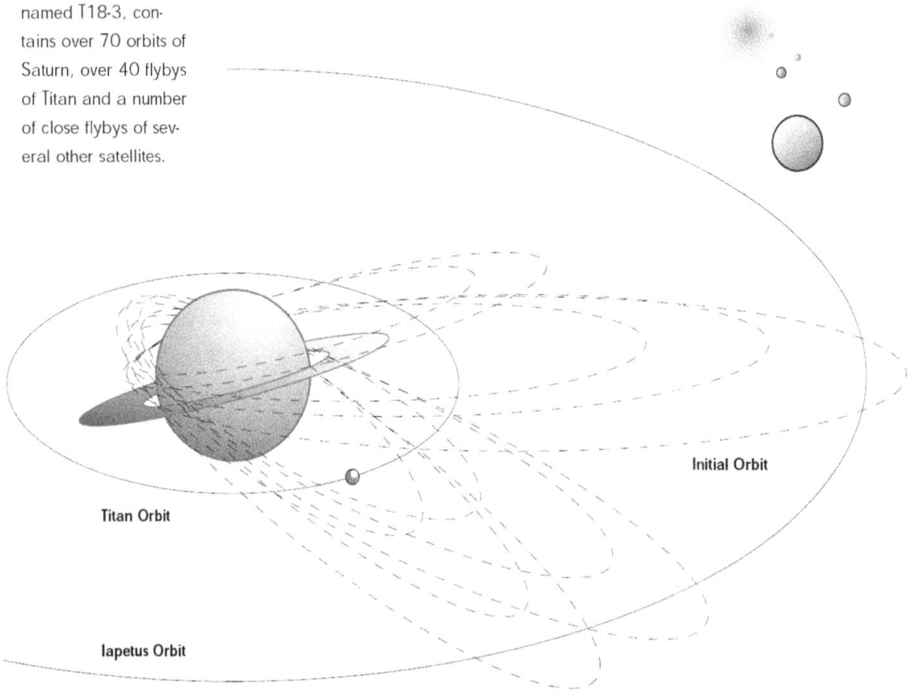

CASSINI–HUYGENS MISSION SCIENCE OBJECTIVES

SATURN

- Determine temperature field, cloud properties and composition of the atmosphere.
- Measure global wind field, including wave and eddy components; observe synoptic cloud features and processes.
- Infer internal structure and rotation of the deep atmosphere.
- Study diurnal variations and magnetic control of ionosphere.
- Provide observational constraints (gas composition, isotope ratios, heat flux) on scenarios for the formation and evolution of Saturn.
- Investigate sources and morphology of Saturn lightning (Saturn electrostatic discharges, lightning whistlers).

TITAN

- Determine abundances of atmospheric constituents (including any noble gases); establish isotope ratios for abundant elements; constrain scenarios of formation and evolution of Titan and its atmosphere.
- Observe vertical and horizontal distributions of trace gases; search for more complex organic molecules; investigate energy sources for atmospheric chemistry; model the photochemistry of the stratosphere; study formation and composition of aerosols.
- Measure winds and global temperatures; investigate cloud physics and general circulation and seasonal effects in Titan's atmosphere; search for lightning discharges.
- Determine physical state, topography and composition of surface; infer internal structure.
- Investigate upper atmosphere, its ionization and its role as a source of neutral and ionized material for the magnetosphere of Saturn.

MAGNETOSPHERE

- Determine the configuration of the nearly axially symmetrical magnetic field and its relation to the modulation of Saturn kilometric radiation.
- Determine current systems, composition, sources and sinks of the magnetosphere's charged particles.
- Investigate wave–particle interactions and dynamics of the dayside magnetosphere and magnetotail of Saturn, and their interactions with solar wind, satellites and rings.
- Study effect of Titan's interaction with solar wind and magnetospheric plasma.
- Investigate interactions of Titan's atmosphere and exosphere with surrounding plasma.

RINGS

- Study configuration of rings and dynamic processes (gravitational, viscous, erosional and electromagnetic) responsible for ring structure.
- Map composition and size distribution of ring material.
- Investigate interrelation of rings and satellites, including embedded satellites.
- Determine dust and meteoroid distribution in ring vicinity.
- Study interactions between rings and Saturn's magnetosphere, ionosphere and atmosphere.

ICY SATELLITES

- Determine general characteristics and geological histories of satellites.
- Define mechanisms of crustal and surface modifications, both external and internal.
- Investigate compositions and distributions of surface materials, particularly dark, organic rich materials and low-melting-point condensed volatiles.
- Constrain models of satellites' bulk compositions and internal structures.
- Investigate interactions with magnetosphere and ring system and possible gas injections into the magnetosphere.

CASSINI—HUYGENS
CHRONOLOGY OF A PLANETARY MISSION

1982

Space Science Committee of the European Science Foundation and the Space Science Board of the National Academy of Sciences form working group to study possible U.S.–European planetary science cooperation. European scientists propose a Saturn orbiter–Titan probe mission to the European Space Agency (ESA), suggesting a collaboration with NASA.

1983

U.S. Solar System Exploration Committee recommends NASA include a Titan probe and a radar mapper in its core program and also consider a Saturn orbiter.

1984–85

Joint ESA–NASA assessment study of a Saturn orbiter–Titan probe mission.

1987

ESA Science Program Committee approves Cassini for Phase A study, with conditional start in 1987.

1987–88

NASA carries out further definition and work on Mariner Mark 2 spacecraft and the missions designed to use it: Cassini and Comet Rendezvous/Asteroid Flyby (CRAF).

Titan probe Phase A study carried out by joint ESA/NASA committee, supported by European industrial consortium led by Marconi Space Systems.

1988

Selection by ESA of Cassini mission Probe to Titan as next science mission; Probe named Huygens.

1989

Funding for CRAF and Cassini approved by U.S. Congress. NASA and ESA release announcements of opportunity to propose scientific investigations for the Saturn orbiter and Titan probe.

1992

Funding cap imposed on CRAF/Cassini: CRAF canceled; Cassini restructured. Launch rescheduled from 1996 to 1997.

1995

U.S. House Appropriations Subcommittee targets Cassini for cancellation, but the action is reversed.

1996

Spacecraft and instruments integration and testing.

APRIL 1997

Cassini spacecraft shipped to Cape Canaveral, Florida.

SUMMER 1997

Final integration and testing.

OCTOBER 1997

Launch from Cape Canaveral, Florida.

APRIL 1998

First Venus gravity-assist flyby.

JUNE 1999

Second Venus gravity-assist flyby.

AUGUST 1999

Earth gravity-assist flyby.

DECEMBER 2000

Jupiter gravity-assist flyby.

DECEMBER 2001

First gravitational-wave experiment.

JUNE 12, 2004

Phoebe flyby; closest approach is 52,000 kilometers.

JULY 1, 2004

Spacecraft arrives at Saturn and goes into orbit around the planet.

NOVEMBER 6, 2004

Release of Huygens Probe on a trajectory to enter Titan's atmosphere.

NOVEMBER 27, 2004

Huygens returns data as it descends through Titan's atmosphere and reaches the surface; Orbiter begins tour of the Saturn system.

JULY 2008

Nominal end of mission.

Cassini–Huygens designed to determine the present state of these bodies and the processes operating on or in them, but it is also equipped to discover the interactions that occur among and between them.

These interactions within the Saturn system are important. An analogy can be drawn to a clock, which has many parts. However, a description of each part and the cataloging of its properties, alone, completely misses the "essence" of a clock. Rather, the essence is in the interactions among the parts. So it is for much of what the Cassini–Huygens mission will be studying in the Saturn system.

It is the ability to do "system science" that sets apart the superbly instrumented spacecraft. The very complex interactions that are in play in systems such as those found at Jupiter and Saturn can only be addressed by such instrument platforms. This is because the phenomena to be studied are often sensitive to a large number of parameters — a measurement might have to take into account simultaneous dependencies on location, time, directions to the Sun and planet, the orbital configurations of certain satellites, magnetic longitude and latitude and solar wind conditions. To deal with such complexity, the right types of instruments must be on the spacecraft to make all the necessary and relevant measurements — and all the measurements must be made essentially at the same time. Identical conditions very seldom, if ever, recur. Thus, it would be totally impossible for a succession, or even a fleet, of "simple" spacecraft to obtain the same result. Furthermore, requiring the instruments to operate simultaneously has a major impact on spacecraft resources such as electrical power. This demand — and the need for a broadly based, diverse collection of instruments — is the reason that the Cassini–Huygens spacecraft is so large.

Both the Huygens Probe and the Cassini Orbiter will study Titan. While the formal set of scientific objectives is the same for both, the Cassini–Huygens mission is designed so that a synergistic effect will be realized when the two sets of measurements are combined. In other words, the total scientific value of the two sets of data together will be maximized because of certain specific objectives for the Probe and the Orbiter. Each time the Orbiter flies by Titan, it will make atmospheric and surface remote-sensing observations that include re-observations along the flight path of the Probe. The Probe's measurements will be a reference set of data for calibrating the Orbiter's observations. In this way, the Probe and Orbiter data together can be used to study the spatial and seasonal variations of the composition and dynamics of the atmosphere.

Launch: Ending and Beginning

Launch is both an ending and a beginning. Nowhere is there a more profound test of the work that has been done, nor a more dramatic statement to mark a shift in program priorities. Nowhere is there more hope — or more tension. A successful launch is everything.

These will be our thoughts as we survey the scene at Cape Canaveral Air Station in Florida. It is October 6, 1997. The launch vehicle is the Titan IVB with two stout Solid Rocket Motor Upgrades (SRMUs) attached. An additional Centaur rocket — the uppermost stage — sits on top of the propulsion stack. This system puts Cassini–Huygens into Earth orbit and then, at the right time, sends it on its interplanetary trajectory.

The "core" Titan vehicle has two stages. The SRMUs are anchored to the first, or lower, stage. These "strap-on" rockets burn solid fuel; the Titan uses liquid fuel. The Centaur is a versatile, high-energy, cryogenic-liquid-fueled upper stage with two multiple-start engines. The Titan IVB/SRMU–Centaur system is capable of placing a 5760-kilogram payload into a geostationary orbit. On top of all this propulsive might sits Cassini–Huygens, protected for its trip through the lower atmosphere by a 20-meter-long payload fairing.

Lift-off is from Cape Canaveral Air Station, launch complex 40. It is early in the morning. The launch sequence begins with the ignition of the two solid-rocket motors, which lift the whole stack off the pad. About 10 seconds after lift-off, the stack continues to accelerate and then it starts to tilt and rotate. The rotation continues until the required azimuth is reached.

At plus two minutes, the first stage of the Titan is ignited, at an altitude of approximately 58,520 meters. A few seconds later, the two solid-rocket motors, now spent, are jettisoned. One and a half minutes pass, and at 109,730 meters, the payload fairing is released.

About five and a half minutes into the flight, the Titan reaches an altitude of 167,330 meters. Here the first stage of the Titan separates and the second stage fires. At launch plus nine minutes, the second stage has burnt out and drops away. Now it is the Centaur's turn to fire. It boosts the remaining rocket–spacecraft stack into a "parking" orbit around Earth and turns off its engines.

Some 16 minutes later, the Centaur ignites for a second time. This burn lasts between seven and eight minutes, and when it is over, the Centaur separates from the spacecraft. The Cassini–Huygens spacecraft is now on an interplanetary trajectory, heading first for Venus, then Venus again, around to Earth, to Jupiter — and at last, Saturn!

CHAPTER 2

The Ringed World

Saturn was the most distant planet that could be seen by ancient astronomers. Like Mercury, Venus, Mars and Jupiter, it appeared to be a bright, star-like wanderer that moved about the fixed stars in the night sky. For this reason, these celestial bodies were called "planetes," the Greek word for "wanderers." Little was learned about Saturn, besides its apparent celestial motions, until the telescope was invented in Holland in the early 1600s.

Historical Observations

In 1609, Galileo Galilei built his own telescope and put it to use in making observations of the night sky. In his initial observations of Saturn, Galileo thought he was seeing three planets because of the poor quality of the lenses of his telescope. By 1614, his notes indicated that he observed the rings for the first time — though he did not identify them as such.

The optical aberrations in early telescopic lenses prevented clear viewing. Because of the way Saturn's rings appeared when viewed through early telescopes, observers spent the early years trying to understand what they interpreted as the planet's odd shape and behavior. Some saw the rings as cup handles. By 1659, Christiaan Huygens had developed the concept of a planetary ring system, which helped the astronomers of the day understand what they were seeing.

Because of its dense cloud cover, all we can see of Saturn is the atmosphere — that is, if we do not look at the rings! Saturn has been a mystery to us since Galileo's first telescopic observations in the early 17th century. Up to the time of the visits made by Voyagers 1 and 2, Saturn appeared as a fuzzy yellow ball with some visible banding and some pole darkening. We now know that the "surface" of Saturn is gaseous and that the patterns present are due to clouds in its gaseous envelope.

From Galileo's time through the next 300 years, telescopes improved and more moons were discovered to be

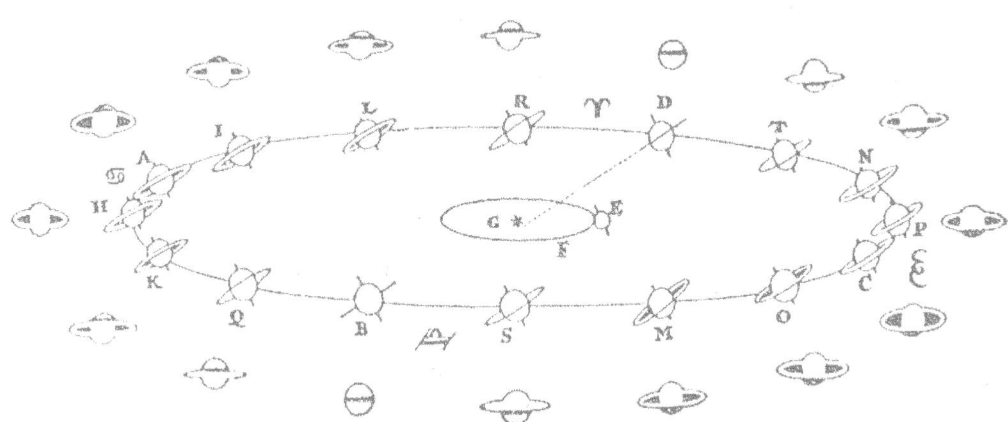

In this diagram from his book, *Systema Saturnium*, Christiaan Huygens explained the inclination of Saturn's rings according to the planet's orbital position with respect to Earth.

orbiting Saturn. But not until the first spacecraft were able to fly by the planet did we really advance our knowledge of Saturn's atmosphere, complicated ring structure, magnetic field, magnetosphere and satellites.

So far, there have been three flyby missions to Saturn. Detailed information about Saturn's structure was obtained by these spacecraft. The first, Pioneer 11, flew by in August 1979. Pioneer's photopolarimeter made important measurements of Saturn's atmosphere, and the data could be assembled into low-resolution images. The atmosphere's appearance changes so rapidly, however, that the time required to take a picture was too long to capture the details. Only subsequent flyby missions by Voyager 1 in October 1980 and Voyager 2 in August 1981 were able to show us the complex structure of the atmosphere and its rapidly changing features. In addition to taking images, the Voyagers conducted many different experiments — the data collected greatly added to our knowledge of Saturn's interior structure, clouds and upper atmosphere.

The composition of planets in the solar system is largely controlled by their temperatures, as determined by their distances from the Sun. Refractory compounds (those with high melting points) were the first to condense, at temperatures around 1500 kelvins, followed by silicates at 1400 kelvins. These substances formed the rocky cores of the planets. While hydrogen is the predominant element in the universe and in our solar system, other gases are present, including water, carbon dioxide and methane, all of which condense below 500 kelvins. As ices, these predominate in the cooler, outer solar system. This gaseous material collected as envelopes around the planets and moons. (Where conditions are right, large quantities of water in its liquid state can form, as exhibited by Earth's oceans, Mars' flood plains and potentially beneath the surface of Europa, a moon of Jupiter.)

The large outer planets contain much of the primordial cloud's gases that were not trapped by the Sun. Hydrogen is the most abundant material in the Sun and in all the large gaseous planets — Jupiter, Saturn, Uranus and Neptune. Each of these giant planets — known as the "gas giants" — has many moons. These moons form satellite systems, suggesting that miniature solar systems formed around the gas giants by processes similar to those that formed the solar system itself.

Characteristics of Saturn

Cool and Slow. Saturn is the sixth planet from the Sun. It is nine-and-a-half times farther from the Sun than Earth. The diameter of the Sun viewed from Saturn is about one-tenth the size of the Sun we see from here on Earth. Sunlight spreads as it travels through space; an area on Earth receives 90 times more sunlight than an equivalent area on Saturn. Because of this fact, the same light-driven photoprocessing in Saturn's atmosphere takes 90 times longer than it would on Earth. You would not have to worry about getting a sunburn on Saturn!

Remembering the astronomer–mathematician Johannes Kepler's laws of planetary motion, the farther away from the Sun, the slower a planet travels in its orbit, and the longer it takes to complete its orbit about the Sun. Saturn travels at an average velocity of only 9.64 kilometers per second, whereas Earth travels at an average velocity of 29.79 kilometers per second. Saturn's yearly orbit about the Sun — a "Saturn year" — is equal to 29.46 Earth years. If you lived on Saturn, you would have only one birthday every 29-plus (Earth) years.

Since the orbit of Saturn is not circular but is elliptical in shape, its distance from the Sun changes during its orbital revolution around the Sun. The elliptical orbit causes a small change in the amount of sunlight that reaches the surface of the planet over the Saturn year, and may affect the planet's upper atmospheric composition over that period.

Slightly Squashed. Saturn's period of rotation around its axis depends on how it is measured. The cloud tops show a rotation period of 10 hours and 15 minutes at the equator, but 23 minutes longer at higher latitudes. A radio signal associated with Saturn's magnetic field shows a period of 10 hours and 39.4 minutes.

The high rotation rate creates a strong centrifugal force, causing an equatorial bulge and a flattening of

This Voyager 1 image shows Saturn's "squashed" appearance — the planet's equatorial bulge and flattened poles.

the planet's poles. As a result, Saturn's equator is 60,330 kilometers from the center, while the poles are only 54,000 kilometers from the center. Almost 10 Earths can be lined up along Saturn's equatorial diameter.

The Mysterious Interior

Low Density. To understand Saturn's interior and evolution, we must apply our basic knowledge of physics to observations of Saturn's volume, mass, gravity, temperature, magnetic field and cloud movement. The clues provided by this body of knowledge are not always easy to decipher, but as we gain more information on Saturn, we are more and more able to relate the pieces of the puzzle.

Saturn has the lowest density of all the planets, because of its vast, distended, hydrogen-rich outer layer. In the early 1900s, we thought that the giant planets might consist entirely of gas. Actually, the giant planets contain cores of heavy elements like iron as well as other components of refractory and silicate compounds, and thus have several times the mass and density of Earth.

Saturn's density is not uniform from its center to the surface — the density in the core is many times that of the surface. An understanding of this distribution is obtained through observations of planetary probes sent from Earth. Observations of the probe trajectories can be used to determine the density distribution throughout Saturn's interior.

The Rocky Core. Using data from the Voyager flybys, planetary scientists have put together a picture of Saturn's interior. We believe that Saturn has a molten rocky core of about the same volume as Earth, but with three or more times the mass of Earth's core. This increased density is due to gravitational compression resulting from the pressure of the liquid and atmospheric layers above the core.

The rocky core is believed to be covered with a thick layer of metallic liquid hydrogen, and beyond that, a layer of molecular liquid hydrogen. The great overall mass of Saturn produces a very strong gravitational field, and at levels just above the core the hydrogen is compressed to a state that is liquid metallic and conducts electricity. (On Earth, liquid hydrogen is usually made by cooling the hydrogen gas to very cold temperatures. On Saturn, liquid hydrogen is very hot, with temperatures of many thousands of kelvins, and is formed under several million times the atmospheric pressure found on Earth.) It is believed that this conductive liquid metallic hydrogen layer, spinning with the rest of the planet, is the source of Saturn's magnetic field — turbulence or convective motion in this layer may be creating the field.

A remarkable characteristic of Saturn's magnetic field is that its axis of rotation is the same as that of the planet. This is different from that of the five other known magnetic fields, of Mercury, Earth, Jupiter, Uranus and Neptune. Present theory suggests that when the axes of rotation and magnetic field are aligned, the magnetic field cannot be maintained.

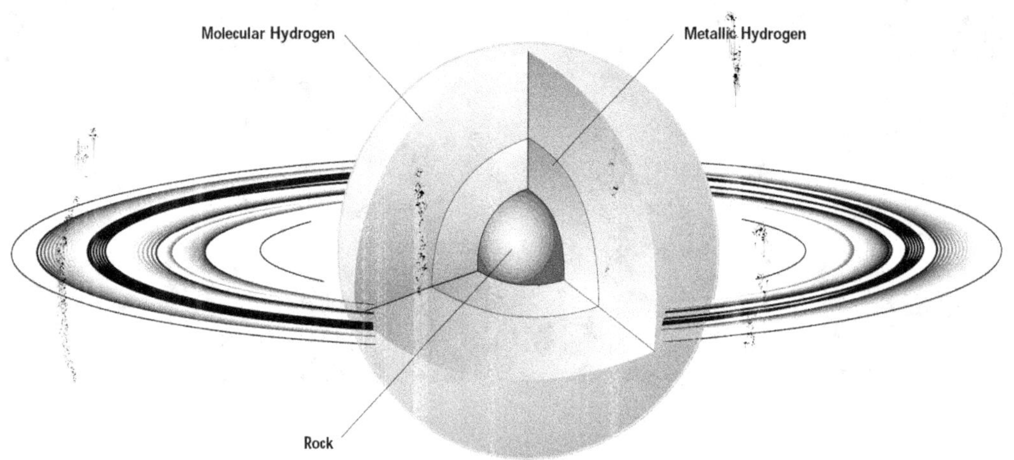

This cutaway diagram shows the basic structure of Saturn's interior.

An Extended Atmosphere

From Liquid to Gas. Above the layer of liquid molecular hydrogen, there is a vast atmosphere that is only surpassed by the atmospheres surrounding Jupiter and our Sun. On Earth, there is a definite separation between the land, the oceans and the atmosphere, but Saturn has only layers of hydrogen that transform gradually from a liquid state deep inside to a gaseous state in the atmosphere, without a well-defined boundary. This is an unusual condition that results from the very high pressures and temperatures found on Saturn.

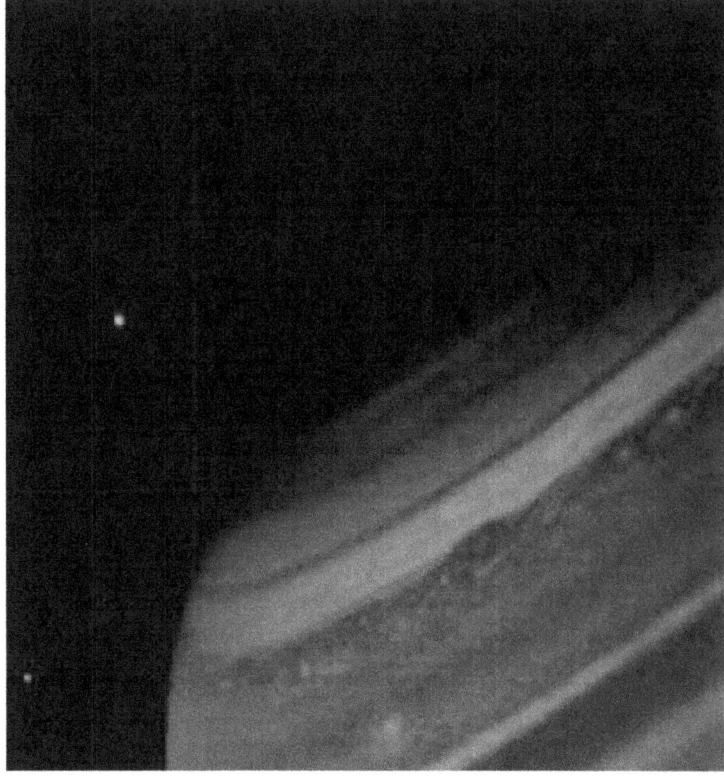

Two satellites orbit far above the raging storm clouds of Saturn in this false-color image taken by Voyager 1.

Because the pressure of the atmosphere is so great, at the point where the separation would be expected to occur, the atmosphere is compressed so much that it actually has a density equal to that of the liquid. This amazing condition is referred to as "supercritical." This can happen to any liquid and gas compressed to a point above critical pressure. Saturn thus lacks a distinct surface, so scientists make measurements from the cloud tops. The reference point is a pressure of one bar (or one Earth atmosphere, 760 millimeters of mercury).

The major component of Saturn's atmosphere is hydrogen gas. If the planet were composed solely of hydrogen, there would not be much of interest to study. However, the composition of Saturn's atmosphere includes six percent helium gas by volume and 0.0001 percent of other trace elements. Using spectroscopic analysis, scientists know that these atmospheric elements can interact to form ammonia, phosphine, methane, ethane, acetylene, methylacetylene and propane. Even a small amount is enough to freeze or liquefy and make clouds of ice or rain possessing a variety of colors and forms.

With the first pictures of Saturn taken by the Voyager spacecraft in 1980, we could see that the clouds and the winds were almost as complex as those found on Jupiter just the year before. There has been an effort to label the belts and zones seen in Saturn's cloud patterns. The banding results from temperature-driven convective flows in the atmosphere, very much the same process that occurs in Earth's atmosphere, but on a grander scale and with a different heat source.

Saturn has different rotation rates in its atmosphere at different latitudes. Differences of 500 meters per second were seen between the equator and nearer the poles, with higher speeds at the equator. This is five times greater than the wind velocities found on Jupiter.

The Exosphere. Like all planets and moons with atmospheres, Saturn's outermost layer of atmosphere is extremely thin. The exosphere is the transition from lower layers to the very tenuous gases and ions within the magnetosphere, which extends out to the moon Titan and beyond. Since its density is so low, the exosphere is easily heated by absorption of sunlight. At the outer edge of the exosphere, the temperature of Saturn's atmosphere is between 400 and 800 kelvins, cooling closer to its base.

The gases in the exosphere are heavily bombarded with light and thus dissociate to the atomic level, that is, all the chemicals in the exosphere are separated into atoms or atomic ions. This material has no way to dissipate the energy absorbed from sunlight except through a rare collision with another atom. Absorbed solar photons heat the exosphere up and the collisions cool it down. How fast the exosphere heats or cools depends upon how far away it is from the bulk of the atmosphere.

The Ionosphere. Lower in the atmosphere, below the exosphere, is the ionosphere. It is characterized by a large abundance of electrons and ions. There are equal numbers of negatively charged electrons and positively charged ions, because the electrons and the ions are formed at the same time. When a high-energy photon of light is absorbed by a neutral or ion, a negative electron is energized to escape, leaving an ion behind with an equal amount of positive charge. The maximum number of charged particles in the ionosphere occurs at a distance of 63,000 kilometers from the planet center, or 3000 kilometers above the one-bar pressure level.

Between the exosphere and the middle of the ionosphere, the predominant ions are H^+, H_2^+, and H_3^+. In the lower ionosphere, where there is an increase in pressure, more complicated hydrocarbon ions predominate. The number of ions present is only a very small percentage of the molecules in the atmosphere, but as in Earth's atmosphere, the ionosphere has an important influence. The ionosphere is like a pair of sunglasses for the planet, filtering out the more energetic photons from sunlight. These energetic photons are the cause of the ionization.

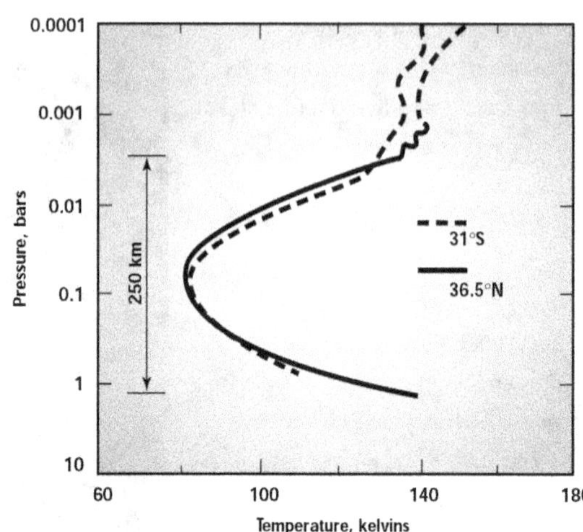

Voyager 2 measurements showed the nearly identical atmospheric structure at two different positions on Saturn.

Cloud Decks. Approaching the planet and from a distance, the Voyager spacecraft could only see a slightly squashed, fuzzy yellow sphere. As the spacecraft drew nearer to Saturn, light and dark bands parallel to the equator appeared. Closer yet, cyclonic storms could be seen all over the planet. Following the paths of the storms, scientists could measure the velocities of the winds. Scientists observed the tops of clouds covering the "surface" of Saturn, which indicated very high wind speeds at the equator and giant storm patterns in bands around the planet. There were holes in the clouds, and more layers below.

A minimum temperature is reached in Saturn's atmosphere at about 250 kilometers above the one-bar level. At this altitude, the temperature is about 82 kelvins. At such low temperatures, the trace gases in the atmosphere turn into liquids and solids, and clouds form. The highest clouds are associated with ammonia ice at the one-bar level, ammonium sulfide at 80 kilometers below this point, and water ice at 260 kilometers below the ammonia clouds. The gases are stirred up from the lower altitudes and condense into ice grains because the temperature is low and the pressure has dropped to about one bar. At this pressure, the condensation process starts and the ammonia, ammonium sulfate and water take the form of ice crystals.

Haze layers are also observed above the temperature minimum, at pressures of 0.1 and 0.01 bar. Both the origin and composition of these hazes are unknown.

Winds and Weather

Saturn is completely covered with clouds. The cloud tops show the effects of the temperature, winds and weather occurring many kilometers below. Hot gases rise. As they rise, they cool and form clouds. As these gas clouds cool, they begin to sink; this convective motion is the source of the billowy clouds we see in Saturn's cloud layer. The cyclonic storms we observe in the planet's cloud tops are much like the smaller versions we see in our daily satellite weather reports on Earth.

The horizontal banding in Saturn's clouds is a result of the different wind speeds at different latitudes. The fastest winds are at the equator; the banding results from the wind shear between different zones of latitude. One possible model suggests that the atmosphere is layered in cylinders that rotate at different rates and whose axes parallel the planet's axis. If this is true, there should be other notable consequences that will be revealed in future observations of Saturn.

There are variations in temperatures on Saturn, as well, which are the driving forces for the winds and thus cloud motion. The lower atmosphere is hotter than the upper atmosphere, driving the vertical motion of gases, and the equator is warmer than the poles because it receives more direct sunlight. Temperature variations combined with the planet's rapid rotation rate drive the horizontal motion of winds in the atmosphere.

A Comparison to Jupiter

Saturn is similar to Jupiter in size, shape, rotational characteristics and moons, but Saturn is less than one-third the mass of Jupiter and is almost twice as far from the Sun. Saturn radiates more heat than it receives from the Sun. This is true of Jupiter as well, but Jupiter's size and cooling rate suggest that it is still warm from the primordial heat generated from condensation during its formation. The smaller Saturn, however, has had time to cool, so some mechanism — such as helium migration to the core

A STAR IS BORN

Studies of star formation indicate that our solar system formed out of a collection of gases and dust, drawn together by gravitational attraction and condensed over millions of years into many stars. The giant gas cloud condensed into rotating pools of higher density in a process called gravitational collapse, because as condensation proceeds, it accelerates. These rotating pools of material condense more rapidly until their temperatures and densities are great enough to form stars. Surrounding each new star, the leftover material flattens into a disk rotating approximately in the plane of the star's equator. This material can eventually form planets — apparently what occurred to form our own solar system. This image, from the Hubble Space Telescope's Wide Field and Planetary Camera 2, shows a part of the Orion Nebula, which is known as a "nursery for young stars." The inset images at far right are possible examples of young stars.

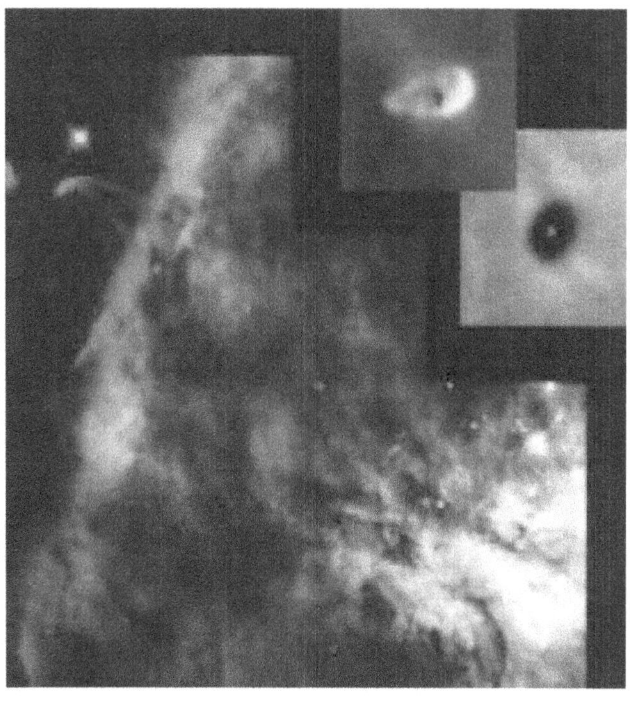

— must be found to explain its continuing radiation of heat.

From Voyager measurements, we learned that Saturn's ratio of helium to molecular hydrogen is 0.06, compared to Jupiter's value of 0.13 (which is closer to the solar abundance and that of the primordial solar nebula). The helium depletion in Saturn's upper atmosphere is believed to be due to helium raining down to the lower altitudes; this supports the concept of helium migration as the heat source in Saturn. Measurements of the planet's energy, radiation and helium abundance will help explain the residual warmth we observe.

Saturn's "surface" features are dominated by atmospheric clouds. They are not as distinct as Jupiter's clouds, primarily because of a haze layer covering the planet that is a result of the weaker insolation from the Sun. This reduced solar radiation leads to greater wind velocities on Saturn. Both Saturn's and Jupiter's weather are driven by heat from below.

Cassini's Global Studies

The Cassini Orbiter's remote-sensing instruments will be the primary data gatherers for global studies of Saturn's atmospheric temperatures, clouds, and composition. Each instrument will provide unique types of data to help solve the puzzles of Saturn's atmosphere.

The Composite Infrared Spectrometer (CIRS), operating in the thermal infrared at very high spectral resolution, is specifically designed to determine temperatures and will spend long periods scanning Saturn's atmosphere. Over the course of a Saturn day, the whole planet will rotate through this instrument's field of view, providing data that will produce a thermal map. Such measurements will help us evaluate the amount of solar heating as compared with the heating generated from the planet's interior. In addition, CIRS can independently measure the temperatures of the different gases composing Saturn's atmosphere. This will give information about how the temperature of Saturn's atmosphere changes at different depths.

Radio science experiments use the spacecraft's radio and ground-based antennas as the science instrument. The Cassini Orbiter will send microwave radio signals through the atmosphere of Saturn to Earth; Saturn's ionosphere and atmosphere will change the signal as it passes through. The radio science "probe" will be used to measure the electron density, temperature, pressure and winds in the ionosphere. Temperature data from radio science experiments depend on the composition of the atmosphere in a unique manner compared with other instruments; thus, the most sensitive determination of helium abundance will be obtained by combining radio science data with measurements from the CIRS.

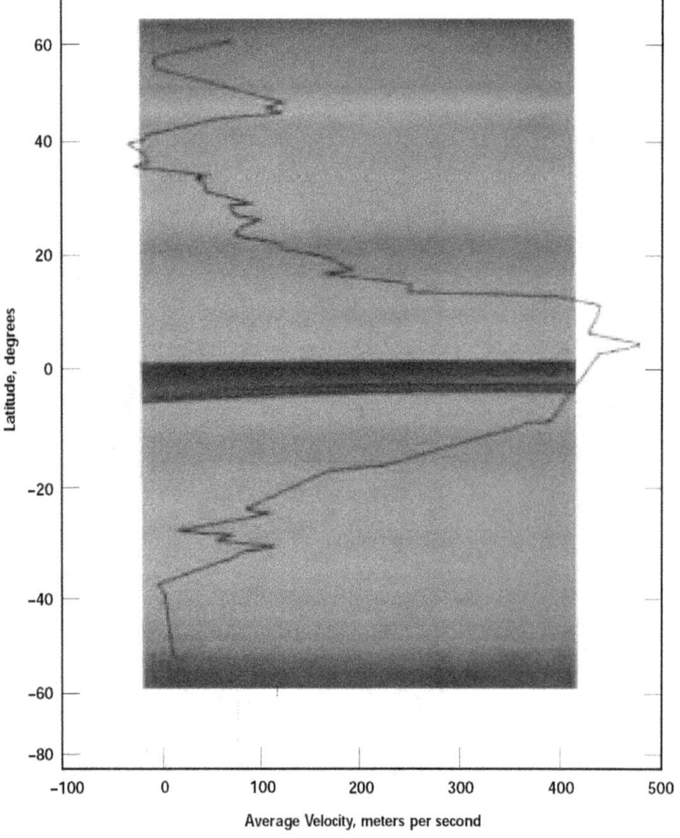

The fastest winds on Saturn are at the equator, here shown surpassing 400 meters per second.

IN THE EYE OF THE BEHOLDER

Taken from Christiaan Huygens' *Systema Saturnium*, these early drawings of Saturn represent the views of: I. Galileo, 1610; II. Scheiner, 1614; III. Riccioli, 1614 or 1643; IV–VII. Hevel; VIII, IX. Riccioli, 1648, 1650; X. Divini, 1646–1648; XI. Fontana, 1636; XII. Biancani, 1616; XIII. Fontana, 1644–1645. Some of the drawings, such as IX, had a very ring-like appearance years before Huygens' theory was accepted.

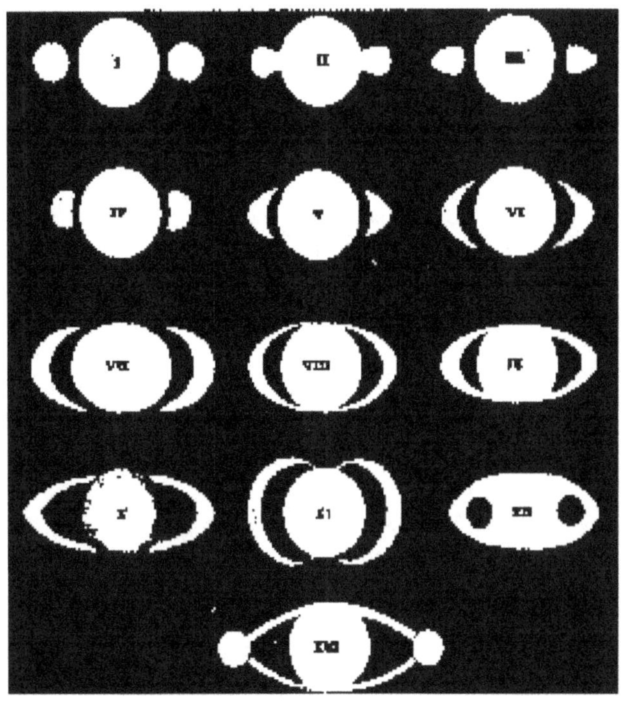

Similar measurements made by the Ultraviolet Imaging Spectrograph (UVIS) and the Visible and Infrared Mapping Spectrometer (VIMS) as the Sun disappears behind Saturn (as seen from the Orbiter) are also very informative. These experiments will observe the atmosphere's variation in composition and opacity with depth by measuring the intensities of various colors of sunlight as the Sun sets into haze and clouds. Each molecule in Saturn's atmosphere can be identified by its unique absorption and emission of the sunlight.

Cassini–Huygens' three spectrometers complement each other in measurements of atmospheric composition. The UVIS and the VIMS measure chemical composition at wavelengths spanning the ultraviolet and near infrared — useful in determining many expected atmospheric components. The CIRS and the UVIS can measure atomic composition as well; the former can delineate the atomic composition of molecules, while the latter can measure atoms directly. The important abundances of deuterium and other interesting isotopes will be investigated to shed light on the origin of our solar system.

The UVIS will follow up on Hubble Space Telescope observations with studies of the energetics of Saturn's aurora. The VIMS and potentially the CIRS can contribute also, and the imaging system will monitor the aurora's changing morphology.

The Imaging Science Subsystem (ISS) will play a major role in understanding the dynamics of Saturn's clouds and weather systems. With wind speeds of 1800 kilometers per hour at the equator, the cloud bands seen beneath the haze layer are subject to considerable turbulence. Indeed, the bands themselves are defined by the winds, separated by zones of high wind shear. In addition, storms can be discerned among the bands, and their dynamics are of considerable interest. Images of the same region made with different filters can be interpreted to indicate the altitudes of various phenomena.

"Saturn-annual" white spots, which appear at approximately 30-year intervals, are not expected during the portion of the cycle when Cassini is studying the planet, but other white spots may fortuitously appear and will be studied by all the optical remote-sensing instruments. Cassini may be able to confirm the theory that the white spots are caused by the upwelling and condensation of ammonia-ice crystals in the atmosphere.

By carefully tracking the Cassini Orbiter's motion around Saturn, data on the deep layers — all the way to the core — of Saturn can be acquired. In these studies, radio signals from the spacecraft will be monitored for changes, especially Doppler shifts of frequency that are different from those predicted from a simple base model. Using the new measurements from the Cassini Orbiter, more detailed models of the internal structure of Saturn will be constructed.

CHAPTER 3

The Mysterious Titan

The exploration of Titan is at the very heart of the Cassini–Huygens mission. As one of the primary scientific interests of the joint NASA–ESA mission, Titan is the sole focus of the Huygens Probe and one of the main targets of the Cassini Orbiter. Through the combined findings from Earth-based telescopes and the Voyager spacecraft, Titan has been revealed to be a complicated world, more similar to a terrestrial planet than a typical outer-planet moon.

A Smoggy Satellite

With a thick, nitrogen-rich atmosphere, possible oceans and a tar-like soil, Titan is thought to harbor organic compounds that may be important in the chain of chemistry that led to life on Earth.

Titan was discovered by Christiaan Huygens about 45 years after Galileo discovered the four large moons of Jupiter. In the early 1900s, the astronomer Comas Sola reported markings that he interpreted as atmospheric clouds. Titan's atmosphere was confirmed in 1944, when Gerard Kuiper passed the sunlight reflecting off Titan through a spectrometer and discovered the presence of methane.

Later observations by the Voyager spacecraft showed that nitrogen was the major constituent of the atmosphere, as on Earth, and established the presence of gaseous methane in concentrations of several percent. The methane participates in sunshine-driven chemistry, which has produced a photochemical smog.

Due to Titan's thick natural smog, Voyager could not see the surface, and instead the images of Titan's disk showed a featureless orange face.

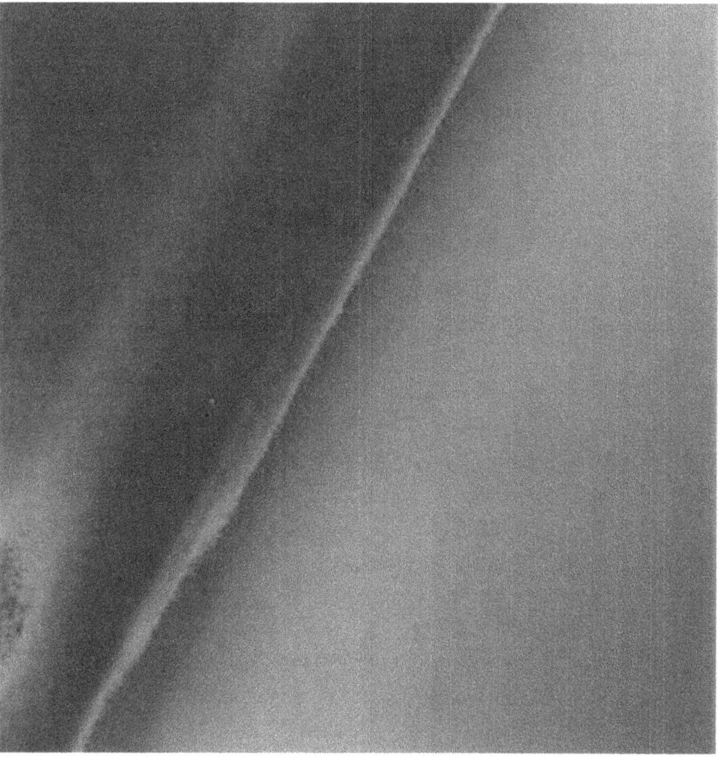

This spectacular view of the edge of Titan was taken by Voyager 1 from a distance of 22,000 kilometers. Tenuous high-altitude haze layers (blue) are visible above the opaque red clouds. The highest of these haze layers is about 500 kilometers above the main cloud deck, and 700 kilometers above Titan's surface.

Spectroscopic observations by Voyager's infrared spectrometer revealed traces of ethane, propane, acetylene and other organic molecules in addition to methane. These organic compounds, known as hydrocarbons, are produced by the interaction of solar ultraviolet light and electrons from Saturn's fast-rotating magnetosphere striking Titan's atmosphere.

Hydrocarbons produced in the atmosphere eventually condense out and rain down on the surface; so Titan may have lakes of ethane and methane, perhaps enclosed in the round bowls of impact craters. Alternatively, liquid ethane and methane may exist in subsurface reservoirs. Titan's hidden surface may have exotic features: mountains sculpted by hydrocarbon rain, rivers, lakes and "waterfalls." Water and ammonia magma from Titan's interior may occasionally erupt, spreading across the surface and creating extraordinary landscapes.

The Cassini–Huygens study of Titan will provide a huge step forward in our understanding of this haze-covered world, and is expected to yield fundamental information on the processes that led to the origin of life on Earth.

By combining the results from the Cassini–Huygens mission with Earth-based astronomical observations, laboratory experiments and computer modeling, scientists hope to answer basic questions regarding the origin and evolution of Titan's atmosphere, the nature of the surface and the structure of its interior. Earth's atmosphere has been significantly altered by the emergence of life. By studying Titan's atmosphere, scientists hope to learn what Earth's atmosphere was like before biological activity began.

The investigation of Titan is divided into inquiries from traditional planetary science disciplines: What is its magnetic environment, what is the atmosphere like, what geological processes are active on the surface and what is the state of its interior? In this chapter, we describe how our knowledge of Titan has developed over the years, focusing on the primary areas of Titan science that the Cassini–Huygens mission will address. The final section discusses how Huygens' and Cassini's instruments will be employed to address fundamental questions, such as: What is the nature of Titan's surface, how have the atmosphere and surface evolved through time and how far has prebiotic chemistry proceeded on Titan?

CASSINI SCIENCE OBJECTIVES AT TITAN

- Determine the abundance of atmospheric constituents (including any noble gases); establish isotope ratios for abundant elements; and constrain scenarios of the formation and evolution of Titan and its atmosphere.
- Observe the vertical and horizontal distributions of trace gases; search for more complex organic molecules; investigate energy sources for atmospheric chemistry; model the photochemistry of the stratosphere; and study the formation and composition of aerosols.
- Measure the winds and global temperatures; investigate cloud physics, general circulation and seasonal effects in Titan's atmosphere; and search for lightning discharges.
- Determine the physical state, topography and composition of the surface; infer the internal structure of the satellite.
- Investigate the upper atmosphere, its ionization and its role as a source of neutral and ionized material for the magnetosphere of Saturn.

Saturn's Magnetosphere

Saturn's magnetic field rotates with the planet, carrying with it a vast population of charged particles called a plasma. (Further discussion of Saturn's magnetosphere can be found in Chapter 6.) The interaction of Saturn's magnetosphere with Titan can be explained by the deflection of Saturn's magnetic field and the ionization and collision of Titan's atmosphere with the charged particles trapped in Saturn's magnetosphere. During the Voyager 1 flyby of Titan, clear indications of changes were observed in the magnetosphere due to the presence of Titan. The signatures included both plasma and plasma-wave effects, along with a draping of Saturn's magnetic field around Titan.

Titan in Saturn's Magnetosphere. Titan orbits Saturn at a distance of about 20.3 R_S (R_S = one Saturn radius). The conditions of the plasma environment in which Titan is submerged can vary substantially, because Titan is sometimes located in the magnetosphere of Saturn and sometimes out in the solar wind. Additionally, the angle between the incident flow and the solar irradiation varies during Titan's orbit. During the Voyager 1 encounter, Titan was found to be within Saturn's magnetosphere and the interaction was described as a transonic flow of plasma (above the local speed of sound).

Scientists expect the characteristics of the incident flow to vary substantially depending on the location of Titan with respect to Saturn's magnetosphere. When Titan is exposed to the solar wind, the interaction may be similar to that of other bodies in the solar system such as Mars, Venus or comets (these bodies have substantial interaction with the solar wind, and, like Titan, have atmospheres but no strong internal magnetic fields).

Titan's Wake. The magnetospheric plasma flowing into and around Titan produces a wake. As with the Galilean satellites at Jupiter, the fast, corotating plasma of Saturn's magnetosphere smashes into Titan from behind, producing a wake that is dragged out in front of the satellite. The process is similar to one in which a wake is created behind a motor boat speeding through the water.

In this analogy, Titan is the motor boat and Saturn's magnetosphere is the water. Instead of the boat going through the water, the water is rushing past the boat. Titan is moving also, in the same direction as Saturn's magnetosphere, but slower, so the magnetospheric plasma actually pushes the wake out in front of the satellite. Scientists expect the wake to be a mixture of plasma from Titan and Saturn's magnetosphere. The wake may be a source of plasma for Saturn's magnetosphere, producing a torus of nitrogen and other elements abundant in Titan's atmosphere.

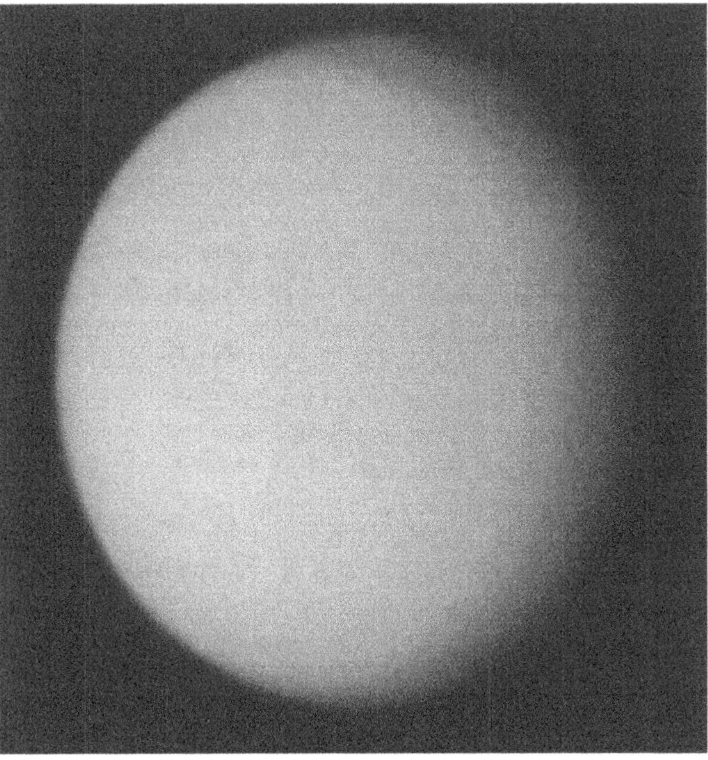

Exploring Titan is like investigating a full-fledged planet. With a radius of 2575 kilometers, Titan is Saturn's largest moon — larger than the planets Mercury and Pluto. Titan is the second largest moon in the solar system, surpassed only by Jupiter's Ganymede. This Voyager image of Titan shows the asymmetry in brightness between the moon's southern and northern hemispheres. Titan's natural smoggy haze blocked Voyager's view of the surface.

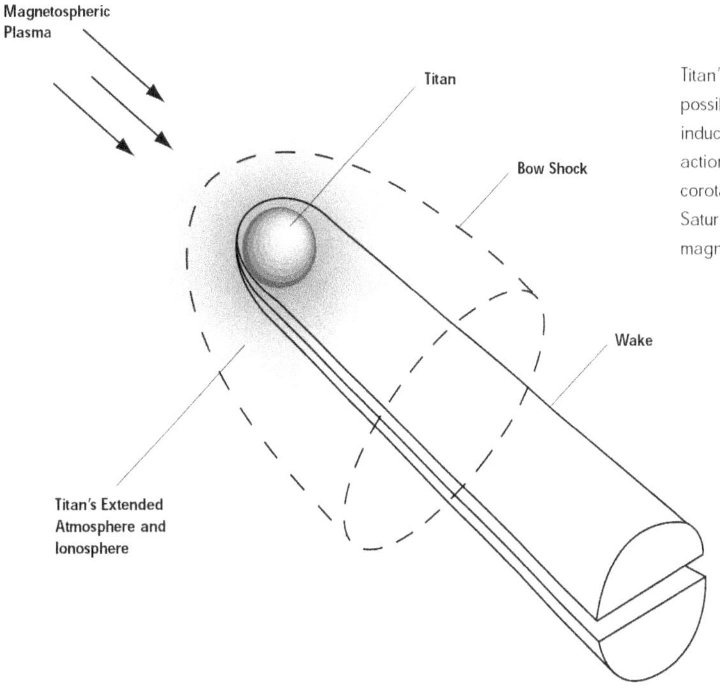

Titan's wake and the possible bow shock induced by the interaction of Titan with corotating plasma in Saturn's vast magnetosphere.

Titan's Torus. The interaction of Titan with Saturn's magnetosphere provides a mechanism for both the magnetospheric plasma to enter Titan's atmosphere and for the atmospheric particles to escape Titan. Voyager results suggested that the interaction produces a torus of neutral particles encircling Saturn, making Titan a potentially important source of plasma to Saturn's magnetosphere. The characteristics of this torus are yet to be explored and will be addressed by the Cassini Orbiter. The interaction of ice particles and dust from Saturn's rings will play a special role as the dust moves out toward Titan's torus and becomes charged by collisions. When the dust is charged, it behaves partially like a neutral particle orbiting Titan, according to Kepler's laws (gravity driven), and partially like a charged particle moving with Saturn's magnetosphere. The interaction of dust with Saturn's magnetosphere will provide scientists with a detailed look at how dust and plasma interact.

Atmosphere–Magnetosphere Interaction. Saturn's magnetospheric plasma, which is trapped in the planet's strong magnetic field, corotates with Saturn, resulting in the plasma flowing into Titan's back side. The flow picks up ions created by the ionization of neutrals from Titan's exosphere and is slowed down while the magnetic field wraps around the satellite. The characteristics of the incident flow are important because the incoming plasma is a substantial source of atmospheric ionization that triggers the creation of organic molecules in Titan's atmosphere. Thus, the aeronomy (study of the physics of atmospheres) of the upper atmosphere and ionosphere is dependent on the plasma flow and the solar radiation as a source of energy. Scientists expect that heavy hydrocarbons are the dominant ions in Titan's ionosphere.

Lightning on Titan? The extensive atmosphere of Titan may host Earth-like electrical storms and lightning. Although no evidence of lightning on Titan has been observed, the Cassini–Huygens mission provides the opportunity to determine whether such lightning exists. In addition to the visual search for lightning, the study of plasma waves in the vicinity of Titan may offer another method. Lightning discharges a broad band of electromagnetic emission, part of which can propagate along magnetic field lines as whistler-mode emission. The emissions are known as "whistlers" because, as detected by radio and plasma-wave instruments, they

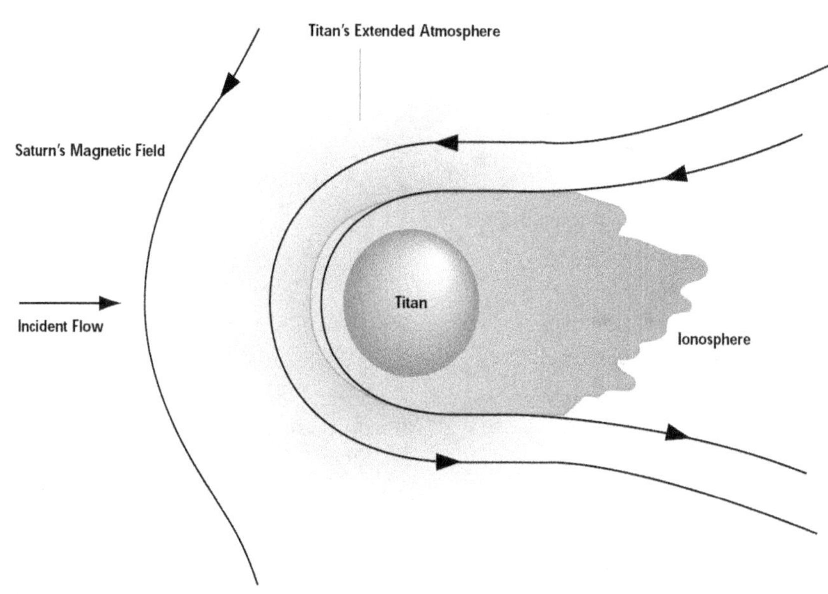

The flow of charged particles in Saturn's magnetosphere past Titan is slowed by the ions created from collisions with Titan's extended atmosphere. Saturn's magnetic field becomes draped around Titan.

SIX GIANT SATELLITES

Satellite (Planet)	Titan (Saturn)	Moon (Earth)	Io (Jupiter)	Europa (Jupiter)	Ganymede (Jupiter)	Callisto (Jupiter)
Distance from Parent, kilometers	1,221,850	384,400	421,600	670,900	1,070,000	1,883,000
Rotation Period, days	15.945	27.322	1.769	3.551	7.155	16.689
Radius, kilometers	2575	1738	1815	1569	2631	2400
Average Density, grams per cubic centimeter	1.88	3.34	3.57	2.97	1.94	1.86

have a signature of a tone decreasing with time (because the high frequencies arrive before the low frequencies). Lightning whistlers have been detected in both the Earth and Jupiter magnetospheres and, besides being detectable from large distances, they offer an opportunity to estimate the frequency of lightning flashes.

Titan's Magnetic Field. The question of whether or not Titan has an internal magnetic field remains open, although Voyager results did not suggest the presence of one. Recent Galileo results from Jupiter indicate the possibility of a magnetic field associated with the moon Ganymede. For Titan there are two possibilities — a magnetic field could be induced from the interaction of Titan's substantial atmosphere with the flow of Saturn's magnetosphere (such as at Venus, with the solar wind), or a magnetic field could be generated internally from dynamo action in a metallic molten core (such as at Earth).

In addition to being important to understanding the Titan interaction with Saturn's magnetosphere, a Titan magnetic field, if generated internally, would place strong constraints on its interior structure.

Titan's Atmosphere

Background. The discovery of methane absorption bands in Titan's spectrum in 1944 was the first confirmation that Titan has an atmosphere. Theoretical analysis followed — would Titan be enshrouded within a warm, atmospheric greenhouse or possess a thin, cold atmosphere with a warm layer at high altitudes? The reddish appearance of Titan's atmosphere led scientists to suggest that atmospheric chemistry driven by ultraviolet sunlight and/or the interaction with Saturn's magnetospheric plasma produced organic molecules. The term "organic" refers here to carbon-based compounds, not necessarily of biological origin.

Later, laboratory experiments and theoretical research were aimed at reproducing the appearance of Titan's spectrum and learning if Titan's atmosphere could indeed be considered a prebiological analog to Earth's atmosphere. Researchers filled laboratory flasks with various mixtures of gases, including methane, and exposed the flasks to ultraviolet radiation. The experiments produced dark, orange–brownish polymers dubbed "tholins," from the Greek word meaning "muddy."

Voyager Results. In 1980 and 1981, the Voyager spacecraft flybys returned a wealth of data about Titan. Voyager 1 skimmed by at a distance of just 4000 kilometers. Voyager images revealed an opaque atmosphere with thin, high hazes. There was a significant difference in brightness between the northern and southern hemispheres and polar hoods, attributed to seasonal variations.

Voyager's solar and Earth occultation data, acquired as the spacecraft passed through Titan's shadow, revealed that the dominant atmospheric constituent is nitrogen. Methane, the atmospheric gas detected from Earth, represents several percent of the composition of the atmosphere. Titan's surface pressure, one and a half bars, is 50 percent greater than Earth's — in spite of Titan's smaller size. The surface temperature was found to be 94 kelvins, indicating that there is little greenhouse warming. The temperature profile in Titan's atmosphere has a shape similar to that of Earth: warmer at the surface, cooling with increasing altitude up to the tropopause at 42 kilometers (70 kelvins), then increasing again in the stratosphere.

The opacity of Titan's atmosphere turned out to be caused by photochemical smog: Voyager's infrared spectrometer detected many minor constituents generated primarily by photochemistry of methane, which produces hydrocarbons such as ethane, acetylene, and propane. Methane also interacts with nitrogen atoms created by the break up of nitrogen to form "nitriles" such as hydrogen cyanide. Titan may well deserve the title "smoggiest world in the solar system."

A person standing on Titan's surface in the daytime would experience a level of daylight equivalent to perhaps 1/1000 the daylight at Earth's surface, given Titan's greater distance from the Sun and the haze and gases blocking the Sun. The light will still be 350 times brighter than moonlight on an Earth night with a full Moon.

In the years following the Voyager mission, a stellar occultation observed from Earth in 1989 provided more data at a Titan season different from the season that Voyager observed. Extensive theoretical and laboratory experiments have also increased our understanding of Titan's complex atmosphere. For instance, this work has helped us to understand complex atmospheric photochemistry and the composition of the haze particles.

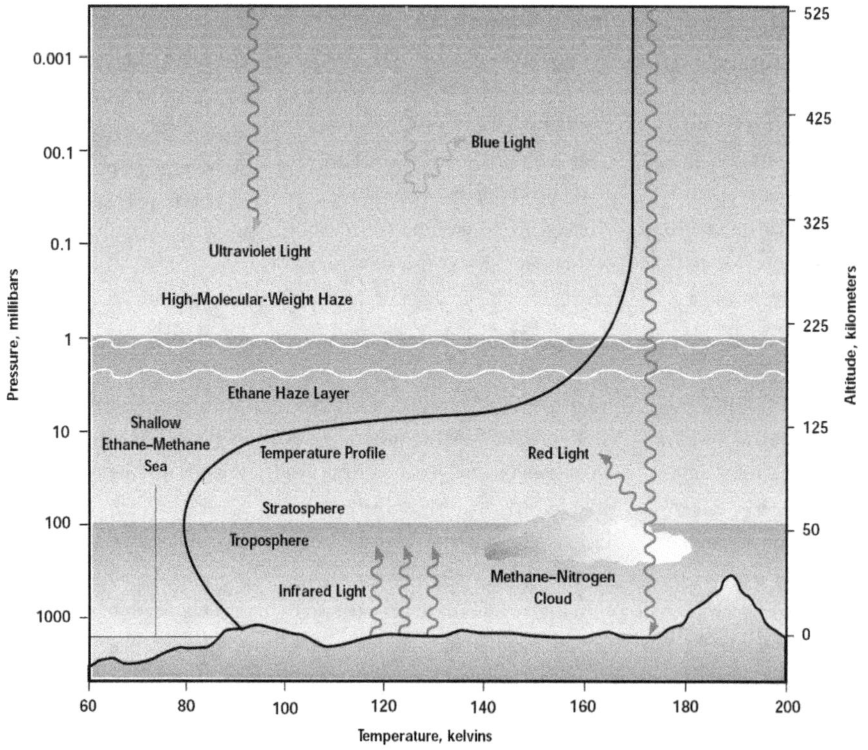

The change in Titan's temperature with altitude is like Earth's: decreasing with increasing altitude up to 50 kilometers, then increasing again above that. This illustration shows atmospheric structure, the location of the surface-obscuring ethane haze and the possible location of clouds.

CONSTITUENTS OF THE TITAN ATMOSPHERE

Chemical Constituent	Common Name	Atmospheric Concentration
N_2	Nitrogen	90–97 percent
Hydrocarbons		
CH_4	Methane	2–10 percent
C_2H_2	Acetylene	2.2 parts per million
C_2H_4	Ethylene	0.1 parts per million
C_2H_6	Ethane	13 parts per million
C_3H_8	Propane	0.7 parts per million
Nitriles		
HCN	Hydrogen cyanide	160 parts per billion
HC_3N	Cyanoacetylene	1.5 parts per billion

To summarize, prior to Cassini, the primary constituents of Titan's atmosphere have been detected, the process by which photolysis of methane has produced a smoggy haze is fairly well understood and the pressure–temperature profile as a function of altitude in the atmosphere has been determined. We do not know the source of the atmosphere, if there is active "weather" (clouds, rain, lightning) nor how the atmosphere circulates. These are the intriguing science questions the Cassini Orbiter and the Huygens Probe will soon investigate.

Atmospheric Origin and Chemistry. What is the source of molecular nitrogen, the primary constituent of Titan's current atmosphere? Is it primordial (accumulated as Titan formed) or was it originally accreted as ammonia, which subsequently broke down to form nitrogen and hydrogen? Or did the nitrogen come from comets? These important questions can be investigated by looking for argon in Titan's atmosphere.

Both argon and nitrogen condense at similar temperatures. If nitrogen from the solar nebula — out of which our solar system formed — was the source of nitrogen on Titan, the ratio of argon to nitrogen in the solar nebula should be preserved on Titan. Such a finding would mean that we have truly found a sample of the "original" planetary atmospheres. Argon is difficult to detect, however, because it is a noble gas — it was not detectable by Voyager instrumentation. The upper limit that has been set observationally is one percent relative to nitrogen; the solar nebula ratio is close to six percent.

Methane is the source of the many other hydrocarbons detected in Titan's atmosphere. It breaks down in sunlight into fragments such as CH_2 and H_2. The CH_2 fragments recombine to produce hydrocarbons. Ethane is the most abundant by-product of the photochemical destruction of methane. The leftover hydrogen escapes from Titan's atmosphere. This is an irreversible process, and the current quantity of methane in Titan's atmosphere — if not replaced — will be exhausted in 10 million years.

The hydrocarbons spend time as the aerosol haze in Titan's atmosphere obscuring the surface. Polymerization can occur at this stage, especially for hydrogen cyanide and acetylene, forming additional aerosols that eventually drift to the surface. Theoretically, the aerosols should accumulate on the surface, and, over the life of the solar system, produce a global ocean of ethane, acetylene, propane and so on, with an average depth of up to one kilometer. A large amount of liquid methane mixed with ethane could theoretically provide an ongoing source of methane in the atmosphere, analogous to the way the oceans on Earth supply water to the atmosphere. Radar and near-infrared data obtained from Earth-based observations show, however, that there is no global liquid ocean, although there could be lakes and seas, or possibly subsurface reservoirs. The ultimate fate of Titan's hydrocarbons, which are expected to exist as liquids or solids on its surface, is a mystery.

Titan has been described as having an environment similar to that of Earth before biological activity forever altered the composition of Earth's atmosphere. It is important to emphasize that a major difference on Titan is the absence of liquid water — absolutely crucial for the origin of life as we know it. Liquid water may occur for short periods after a large comet or meteoroid impact, leading perhaps to some interesting prebiotic chemistry. However, the surface temperatures at Titan are almost certainly cold enough to preclude any biological activity at Titan today.

Winds and Weather. Titan's methane may play a role analogous to that of water on Earth. Might we expect Titan to have methane clouds and rain? At this point, the data suggest otherwise. Voyager infrared data are "best fit" with an atmosphere that is supersaturated with methane. This means that the methane wants to form rain and snow but lacks the dust particles on which the methane could condense.

Titan appears to have winds. The temperature difference from the equator to 60 degrees latitude may be as much as 15 kelvins, which suggests that Titan might have jet streams similar to those in Earth's stratosphere. Wind speeds in Titan's stratosphere may reach 100 meters per second. The occultation of the star 28 Sgr observed from Earth in 1989 confirmed this theoretical analysis by detecting the shape of Titan's atmospheric bulge, which is influenced by high-altitude winds. In the troposphere, the temperature as a function of latitude varies by only a few degrees, and the atmosphere should be much calmer.

Titan's Surface

The surface of Titan was not visible through the limited spectral range of the Voyager cameras. Our knowledge of the surface of Titan comes from much more recent Earth-based images, acquired at longer wavelengths with the Wide Field and Planetary Camera aboard the Hubble Space Telescope. Brightness variations are evident, including a large, continent-size region on Titan's surface with a distinctly higher albedo (reflectance) at both visible and near-infrared wavelengths.

Preliminary studies suggest that a simple plateau or elevation difference on Titan's surface cannot explain the image features, and that the brightness differences must be partly due to a different composition and/or roughness of material. Like other moons in the outer solar system, Titan is expected to have a predominantly water-ice crust. Water at the temperatures in the outer solar system is as solid and strong as rock. Observations show weak spectral features indicative of ice on Titan's surface, but some dark substance is also present. This suggests that something on the surface is masking the water ice.

Surface Geology. At the resolution provided by the Hubble Space Telescope, we can establish Titan's rotation rate, and also agree that this moon has a continent-size albedo feature. The surfaces of solid bodies in the solar system have been altered primarily by three processes: impact cratering, volcanism and tectonics. Erosion may also be important on bodies with atmospheres. By studying the surface of a body, scientists can determine how it has evolved — when the surface solidified, the subsequent geological processes and how the surface and atmosphere interact.

Planetary geologists use crater statistics to determine the relative age of a surface. Since the population of im-

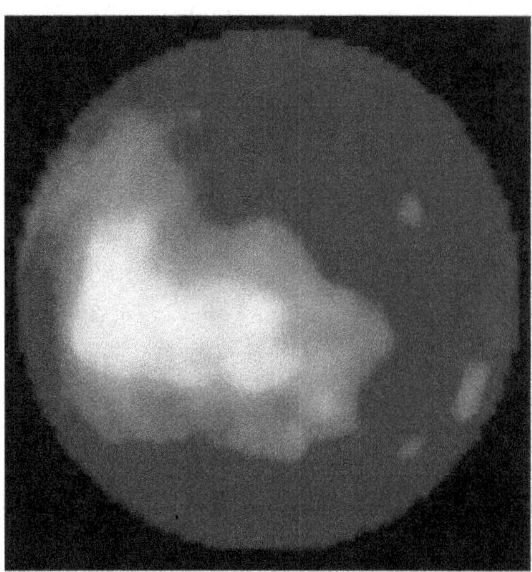

A composite of images of Titan's surface acquired in 1994 by the Hubble Space Telescope shows brightness variations, including a large region that appears bright at visible and near-infrared wavelengths. This region is also a good reflector of radar.

The processes of plate tectonics and erosion have erased the signature of all but the most recent of Earth's impact craters, such as the Meteor Crater in Arizona.

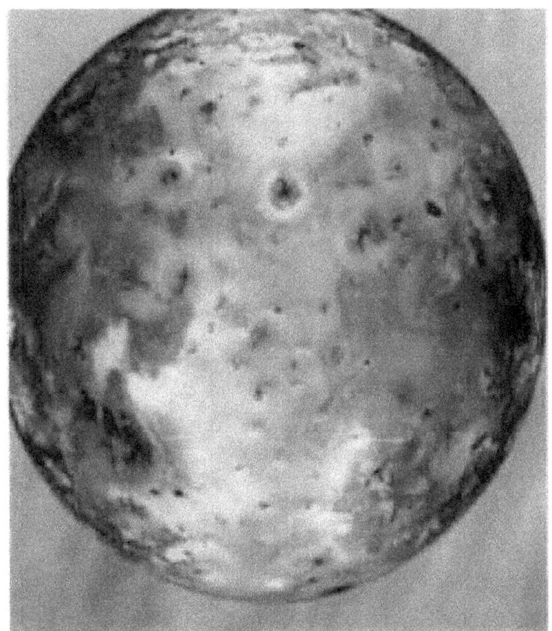

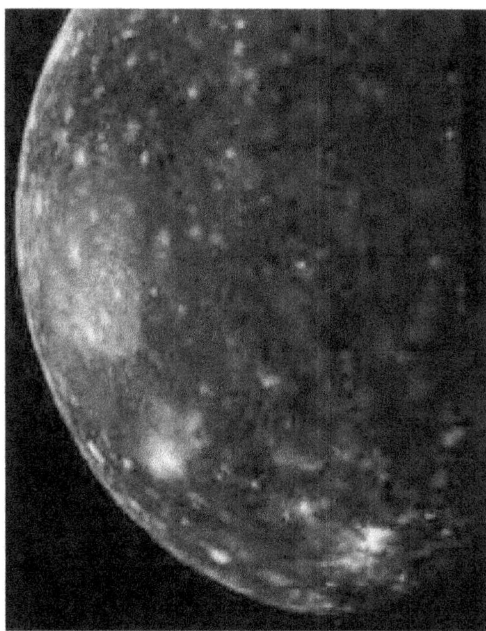

By contrast, in the case of Jupiter's moons, Callisto's cratered surface froze early in the history of the solar system, and has experienced only limited evolution since that time. Continuous volcanism gives Io (left) the youngest surface in the solar system.

pactors was greatest early in the history of the solar system, the more craters there are on a surface, the older the surface must be. Looking at a body like Jupiter's moon Callisto, covered with impact craters, it is clear that the surface is old and that the effect of geological processes has been minimal.

On Earth, tectonic forces and erosion have obliterated all old craters and erosion has modified the few recent ones. On a body like Io, another moon of Jupiter, active volcanism has covered all signs of craters, and the young landscape is dominated by volcanic features. At which end of this spectrum will we find Titan?

Earth-based radar data suggest a surface similar to that of Callisto. Titan's size alone suggests that it may have a surface similar to Jupiter's moon Ganymede — somewhat modified by ice tectonics, but substantially cratered and old. If Titan's tectonic activity is no more extensive than that of Ganymede, circular crater basins may provide storage for lakes of liquid hydrocarbons. Impactors create a layer of broken, porous surface materials, termed "regolith," which may extend to depths from one to three kilometers. The regolith could provide subsurface storage for liquid hydrocarbons as well.

In contrast to Ganymede, Titan may have incorporated as much as 15 percent ammonia as it formed in the colder, Saturn region of the solar nebula. As Titan's water-ice surface froze, ammonia–water liquid would have been forced below the surface. This liquid will be buoyant relative to the surface water-ice crust; thus, ammonia–water magma may have forced its way along cracks to the surface, forming exotic surface features.

Jupiter's moon Ganymede is the only moon in the solar system larger than Titan. The two moons may have similar geology — with regions modified by ice tectonics, but a surface that basically solidified three to four billion years ago, now scarred by craters. This view of Ganymede shows a Galileo image superimposed on a much less detailed Voyager image.

What amount of weathering of the surface might Cassini–Huygens see? On Earth, water accomplishes weathering because as it expands in its freeze–thaw cycle, rocks are broken up. In its liquid phase, water acts as the medium for many chemical reactions. On Titan, condensed hydrocarbons may or may not participate in a weathering process, but may also be expected to form a solid and/or liquid veneer over the icy surface.

The high-contrast features seen in the Hubble Space Telescope images are not consistent with a surface uniformly covered with liquid, suggesting some transport of the hydrocarbons into lakes or subsurface reservoirs.

Interior Structure

Titan's average density is 1.88 grams per cubic centimeter, suggesting a mixture of roughly 50 percent rocky silicate material and 50 percent water ice. Given the temperature and pressure of the solar nebula at Saturn's distance from the Sun when Titan was accreting, it is possible that methane and ammonia would have been mixed with the water ice. The formation of Titan by accretion was undoubtedly at temperatures warm enough for Titan to "differentiate," that is, for the rocky material to separate to form a dense core, with a water–ammonia–methane ice mantle.

The mixture of ammonia with water could ensure that Titan's interior is still partially unsolidified, as the ammonia will effectively act like antifreeze. Radioactive decay in the rocky material in the silicate-rich core would heat the core and mantle for approximately three billion years, further strengthening the case that a liquid layer in the mantle could exist to the present. Titan's solid water-ice crust may be laced with methane clathrate — methane trapped in the structure of water ice. This could provide a possible long-term source for the methane in Titan's atmosphere — if the methane were freed by ongoing volcanism. An alternative model for providing the methane is a porous crust where the methane–water clathrate ratio is approximately 0.1.

Cassini–Huygens Experiments

On the first orbit after Cassini has executed its Saturn orbit insertion maneuver, the Huygens Probe will be targeted for Titan. Probe data are obtained during the descent through Titan's atmosphere and relayed to the Orbiter; then the Orbiter mission continues with over 40 more Titan flybys.

Magnetospheric Investigations. As the Orbiter encounters Titan, the science instruments will work together to investigate the fields and particles science associated with Titan and help determine if the satellite has an intrinsic magnetic field. The Cassini Plasma Spectrometer will map out the plasma flow. The Magnetospheric Imaging Instrument will measure high-energy particles and image the distribution of neutral particles surrounding Titan during the approach and departure of the Orbiter.

Through the measurement of plasma waves, the Radio and Plasma Wave Science instrument will investigate the interaction between the plasmas and high-energy particles, and will also search for radio emission associated with Titan. The Cosmic Dust Analyzer will map out the dust distribution. The Ion and Neutral Mass Spectrometer and the Cassini Plasma Spectrometer will measure the composition of Titan's atmosphere and exosphere directly during the flyby, to determine composition and study the interaction with Saturn's magnetosphere.

Atmospheric Investigations. The main purpose of the Huygens Probe is to return in situ measurements as it descends through Titan's atmosphere and lands on the surface. The Probe's

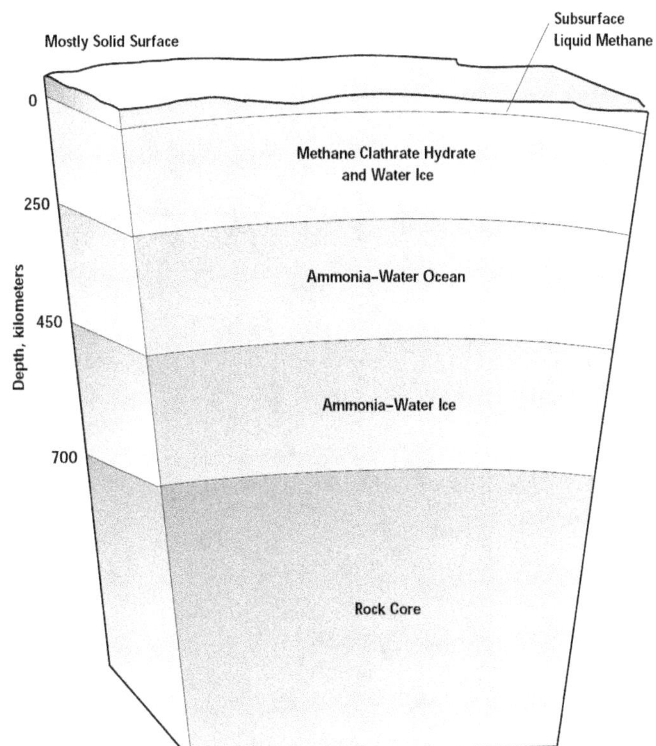

This model of Titan's interior envisions an underground ocean of liquid ammonia and water, 250 kilometers beneath the solid surface.

suite of instruments has been optimized for the return of information on critical atmospheric parameters. The atmospheric structure instrument will measure temperature, density and pressure over a range of altitudes, and also measure the electrical properties of the atmosphere and detect lightning.

The Aerosol Collector and Pyrolyser will vaporize aerosols, then direct the resulting gas to the Gas Chromatograph and Mass Spectrometer, which will provide a detailed inventory of the major and minor atmospheric species, including organic molecules and noble gases — in particular argon. The instrument for the Doppler Wind Experiment will track the course of the descending Probe, giving wind direction and speed at various altitudes in Titan's atmosphere. The Descent Imager and Spectral Radiometer will take pictures and record spectra of the clouds and surface. Measurements of the Sun's aureole collected by this instrument will help scientists deduce the physical properties of the aerosols (haze and cloud particles) in Titan's atmosphere. The instrument will also acquire spectra of the atmosphere at visible and near-infrared wavelengths and measure the solar energy deposited at each altitude level of the atmosphere.

The data acquired by the Probe will be compared with the continuing coverage provided by the Orbiter's subsequent flybys. The Orbiter's three spectrometers (the Ultraviolet Imaging Spectrograph, the Visible and Infrared Mapping Spectrometer and the Composite Infrared Spectrometer) will detect chemical species at a variety of levels in the atmosphere. By mapping the distribution of hydrocarbons as a function of time, the information from the spectrometers will address the sources, sinks and efficiency of the photochemical processes making up the hydrocarbon cycle. The Ultraviolet Imaging Spectrograph can also detect argon and establish Titan's deuterium–hydrogen ratio.

The Orbiter will study the circulation of Titan's atmosphere using the Visible and Infrared Mapping Spectrometer and the Imaging Science Subsystem's cameras to track clouds. For the study of Titan's weather, scientists will combine data from the Orbiter's Composite Infrared Spectrometer and the Radio Science Instrument experiments to develop temperature profiles and thermal maps.

Titan's subsurface structure may be key to understanding the properties of its surface. Photochemical models of Titan's atmosphere predict the existence of a global ocean of liquid ethane and methane, acetylene and other hydrocarbon precipitates — while Earth-based radar data and near-infrared images suggest a solid surface. If Titan's crust is sufficiently porous, storage space could exist for large quantities of liquid hydrocarbons, which would not show up in the radar data. The porous regolith produced by impacts could be over one kilometer thick, providing sufficient volume to hide the expected hydrocarbons.

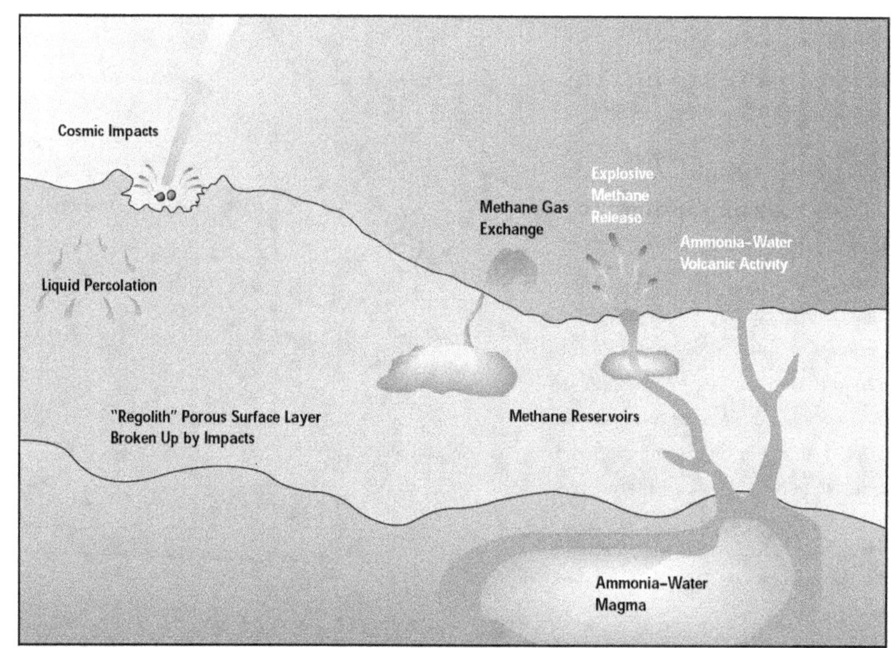

Surface Science. The highest resolution, clearest images of Titan's surface will come from the Descent Imager and Spectral Radiometer on the Huygens Probe. This camera will take over 500 images under the haze layer — where the atmosphere may be very clear — during the Probe's journey to the surface.

If the Probe survives a landing on liquid, aerosol drifts or ice, chemical samples may be analyzed using the gas chemical analyzer. The camera will image the surface, while information on the atmosphere will be relayed by the Huygens Atmospheric Structure Instrument and Doppler Wind Experiment instrument. The Probe's surface science instrumentation will measure deceleration upon impact, test for liquid surface (if waves are present, the instrument can detect the bobbing motion) or fluffy surface and conduct investigations of liquid density, optical refractivity and thermal and electrical properties.

Artist's conception of the Huygens Probe descending through Titan's atmosphere.

This artist's rendering illustrates the Cassini Orbiter's radar as it maps swaths of Titan's surface.

The Cassini Orbiter instruments have the capability to penetrate the hazy atmosphere for data gathering. The Orbiter's radar can map the surface through the clouds with high-resolution swaths in a manner similar to the Magellan mission's mapping of Venus' cloud-enshrouded surface. The radar will provide data on the extent of liquid versus dry land and map surface features at high resolution (hundreds of meters) in narrow swaths.

The Orbiter's camera and visible and infrared radiometer can provide global coverage by imaging through long-wavelength "windows," where atmospheric aerosols do not block sunlight. The combined set of data from these instruments will allow scientists to determine the dominant geological processes — cratering, volcanism and erosion. Titan's spin vector and reference features for making maps will be established, and data from the radar altimeter will provide estimates of the relative elevations. If large bodies of liquid exist, winds may be detected through measurements of the surface roughness.

Interior Structure. Titan's interior structure can be determined indirectly. As it passes close to Titan, the Cassini Orbiter's path will be altered by the satellite's gravitational field. The resulting change in the Orbiter's velocity will be detected on Earth as a Doppler shift in the spacecraft's radio signal. Understanding the basic structure of Titan's gravitational field allows scientists to determine the satellite's moment of inertia, thereby establishing whether Titan has a differentiated core and layered mantle.

Titan's eccentric orbit carries it to varying distances from Saturn. As Titan gets nearer and farther from Saturn, the shape of the satellite changes due to tidal forces. This may cause a detectable change in Titan's gravitational field that can be measured by flying close to Titan a number of times when the satellite is at different locations in its orbit. Such deformation is a measure of the rigidity of Titan, as it flexes because of tidal forces, which can then be used to infer whether or not it has a liquid layer in the mantle.

A World of Its Own

The interaction between Titan's atmosphere and magnetosphere is very complex and many questions still remain. Most of the progress to date has been made using Voyager 1 results, combined with computer modeling. Questions concerning precise interaction, such as the importance of ionization, neutral particle escape, shape of the Titan wake and how the interaction varies over the Titan orbit could all soon be answered by the Cassini–Huygens mission.

With over 40 flybys planned, the Cassini Orbiter should be able to answer the question of whether or not Titan has a significant internal magnetic field.

Understanding Titan's methane–hydrocarbon cycle is a major goal for Cassini–Huygens. Investigating the circulation of the atmosphere and the transport of these hydrocarbons is important to understanding Titan's climate and weather in many time domains: over days, over seasons and over the age of the solar system. The source of atmospheric methane is an intriguing puzzle — methane in Titan's atmosphere will be depleted by irreversible photolysis in less than 10 million years, so there needs to be a replenishing source of methane at or near the surface.

Titan presents an environment that appears to be unique in the solar system, with a thick, hazy atmosphere; a possible ocean of hydrocarbons; and a surface coated with precipitates from the atmosphere — similar to the scenario that scientists believe led to the origin of life on Earth. In the three centuries since the discovery of Titan, we have come to see it as a world strangely similar to our own — yet located about one and a half billion kilometers from the Sun.

When the Cassini Orbiter and Huygens Probe arrive at Titan, they will provide us with our first close-up view of Saturn's largest satellite. With the Probe's slow descent through Titan's atmosphere — with images of the surface and chemical composition data — and the Orbiter's radar maps of the moon's surface, we will once again be shown another world.

CHAPTER 4

Those Magnificent Rings

Galileo first pointed his tiny telescope at Saturn in July of 1610. Since then, the rings of the giant planet have become one of the great, enduring mysteries of our solar system. Even today, those magnificent rings are the most requested object for viewing at any public telescope session. While the beauty of the rings is universally known, the mystery of their existence has challenged the explanations of astronomers for four centuries.

Historical Background

For reasons that are lost to us, Galileo initially paid little attention to Saturn beyond his declaration that it was a triple planet. To him and other early viewers, the nature of the rings was masked by the poor quality of the telescopes of the time. After his initial observations, Galileo seemed to lose interest in Saturn and did not view the planet again for two years.

This enhanced Voyager 2 image shows some hints of possible variations in the chemical composition of Saturn's rings.

When Galileo again chanced to look at Saturn, in 1612, the planet's appearance had totally changed. Saturn no longer seemed "triple," but appeared as a single "ball," similar to the appearance of Jupiter in his small instrument. The puzzle of Saturn, which would grow for centuries, had begun to take shape.

In 1616, when Galileo once again returned to his observations of Saturn, the planet had changed again! The features he had initially seen as separate bodies now had an entirely different appearance; he termed them "ansae" or "handles."

Ring Nature

Once the changing appearance of Saturn had been explained by a ring around the planet, the next question

HUYGENS' HANDLE ON THE RINGS

Following Galileo's characterization of Saturn's "handles," various observations were made by astronomers until 1655, when Christiaan Huygens turned a telescope of far better quality than Galileo's toward the planet and discovered its large moon, Titan.

Within a year, Huygens had arrived at a theory to explain some of the mystery regarding the continually changing appearance of Saturn – he correctly asserted that Saturn was surrounded by a disk or ring whose appearance varied because of the inclination of Saturn's equatorial plane and the planet's orbital motion around the Sun. Shown here is the model of Saturn and its rings used by the Accademia del Cimento to test Huygens' hypothesis.

Huygens had some details about the rings wrong, but his overall theory was a concise explanation of the ring phenomenon and was soon widely accepted. By 1671, when Huygens' theory correctly predicted the rings' "disappearance," his model was accepted as fact – he indeed had grasped the "handle" on the rings.

to occupy astronomers concerned the nature of the ring, or rings. Were the rings solid, or were they a swarm of tiny satellites in orbit, as suggested in 1660 by Christiaan Huygens' close friend, Jean Chapelain? The accounts of various observers caused adjustments to Huygens' theory, and the observations of Jean-Dominique Cassini added information to the scientific debate.

Cassini noted that the brighter and dimmer portions of the rings were separated by a dark band, which he interpreted to be a "gap." This meant that there were actually two rings, but the thought of two solid spinning disks around Saturn was a bit much for astronomers to believe. The "solid disk" theory began to lose its appeal. Instead, by the beginning of the 18th century, the notion that Saturn's rings were actually a swarm of tiny, orbiting satellites gained wide favor in the astronomical community.

During the 19th century, as telescope quality improved, astronomers made numerous studies of Saturn's rings. The quality of observations also generally improved — the matter of the rings' nature was being addressed. In 1857, James Clerk Maxwell put forth a now-famous theory about the nature of the rings, attributing them to swarms of particles and providing a good explanation for the existence of such a system. In 1866, Daniel Kirkwood noted that the Cassini division (or gap) lay at an orbital resonance point[1] with the orbit of Mimas; he later noted that the Encke division also lay at a resonance point.

In 1872, James Keeler noted a narrow gap in the outer A-ring, termed by later observers as the Keeler gap[2]. In 1889, E. E. Barnard made measurements of an occultation of Iapetus by the translucent C-ring, finding it to be a semitransparent ring of nonuniform optical density. In 1895, Keeler proved using spectroscopic evidence that the rings were composed of swarms of particles moving in Keplerian orbits, thus validating Maxwell's theory. In 1908, Barnard made a series of observations at the time of Earth's passage through Saturn's ring plane, concluding that the Cassini division was not devoid of particles.

Over the years, interesting features noted by Earth-based astronomers were not readily subject to verification. In fact, many observations about Saturn remained in dispute until the first robotic probes — Pioneer 11, Voyager 1 and Voyager 2 — flew close to the planet. Viewing Saturn from Earth involves peering through a turbulent atmosphere. Visual acuity varies from person to person: what one observer glimpses for a few seconds now and then might never be seen by another observer at a less favorable observing site or with less acute eyesight.

Even the advent of photography in the late 19th century did not improve discovery verification, since photographic studies were more subject to the woes of looking through Earth's turbulent atmosphere. Only the passes by Pioneer and Voyager

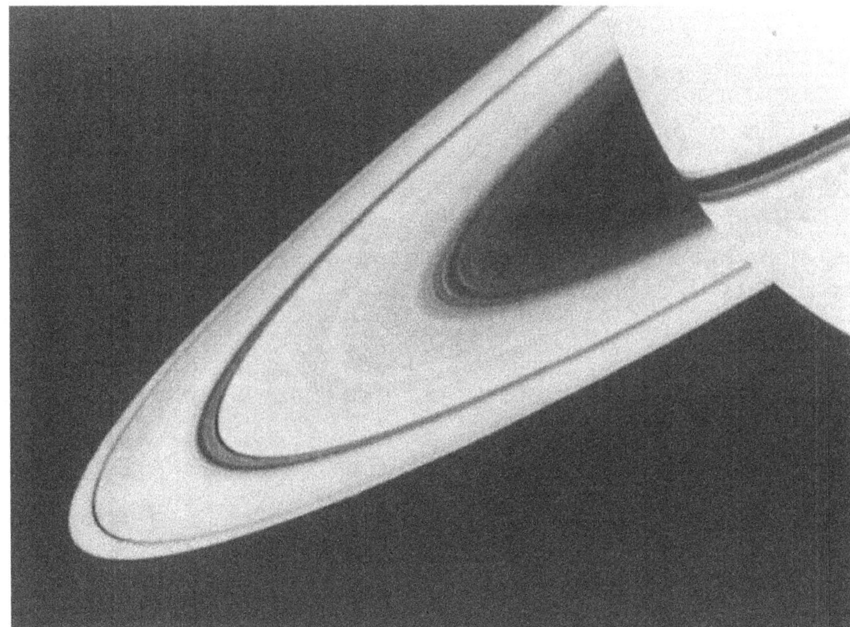

Voyager 1 captured this striking image of Saturn and its rings, with the rings cutting a dark shadow across the planet.

This false-color image of Saturn's outer A-ring was constructed from green, violet and ultraviolet frames. Voyager 2 captured the image on August 23, 1981.

crystallized centuries of Saturn observations, almost instantly sorting observational fact from fiction — just as the Mariner flybys shattered the myths of Martian canals and civilizations.

Ring Structure

Large-Scale Features. Viewing Saturn and its rings through a modest telescope easily allows us to see that the ring system is divided into several visually different radial zones. The outer zone (visible to an Earth observer) is known as the A-ring, while the brighter, inner zone is known as the B-ring. Between these two zones we can sometimes see the gap known as the Cassini division; on an exceptional viewing night, we may see an inner, semitransparent ring, the C or Crepe ring.

This view of the rings, while modest, nevertheless introduces us to the system's unexplained features. Why are there several morphologically distinct rings? Beyond the A, B, C and E-rings, which can be seen or imaged from Earth-bound telescopes, Saturn has several other distinct rings — the D, F and G-rings — detected in the images from Pioneer and Voyager.

The mysteries of Saturn's rings lie at an even more fundamental level: Why are the rings there at all? How did the rings form? How stable is the ring system? How does the system maintain itself? So far, we have only bits and pieces of answers and much speculation about these questions.

Smaller-Scale Features. As Pioneer and Voyager moved closer to Saturn, the spacecraft captured images that made clear the features glimpsed over the centuries, as well as many previously unseen features. Pioneer found a new ring feature (the F-ring) outside the A-ring, which became an object of major interest when Voyager later arrived. The fields and particles sensors on Pioneer 11 showed the possible presence of satellites in the vicinity of the new F-ring. This ring proved to be particularly interesting — it is a narrow ring, and the satellites that (as we now know) maintain the ring in position were the first proof of a theory put forth in 1979 to explain the narrow rings of Uranus.

THE RINGS OF SATURN

Ring	Distance, kilometers*	Width, kilometers
D	66,970	7,500
C	74,500	17,500
B	92,000	25,400
A	122,170	14,610
F	140,180	50
G	170,180	8,000
E	180,000	300,000

** Distance from Saturn to closest edge of ring.*

SATURN'S FLYING SNOWBALLS

Saturn's rings have intrigued scientists on Earth for nearly four centuries. Seen up close, the rings may be even more astonishing than they appear at a distance. Although the rings stretch over 40,000 kilometers in width, they may be as little as 100 meters in thickness. The ring particles — from a few micrometers (one micrometer = one millionth of a meter) to a few meters in size — have been described as ice-covered rocks or icy snowballs. Scientists are continuing in their attempts to analyze ring composition, hoping that the Cassini–Huygens mission will bring them clearer clues to the icy puzzle.

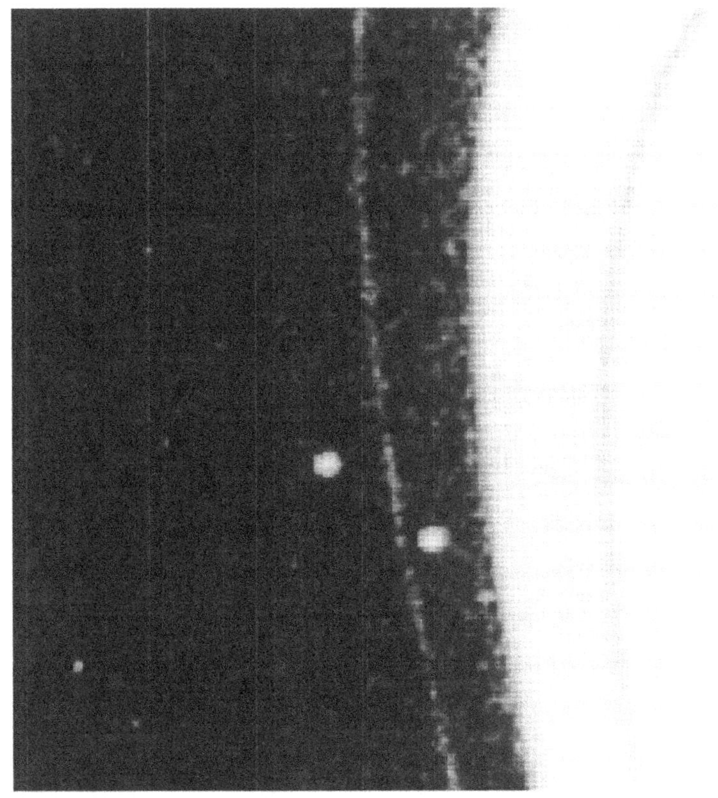

Prometheus and Pandora, two tiny satellites, shepherd Saturn's F-ring, which is multistranded and kinked in places.

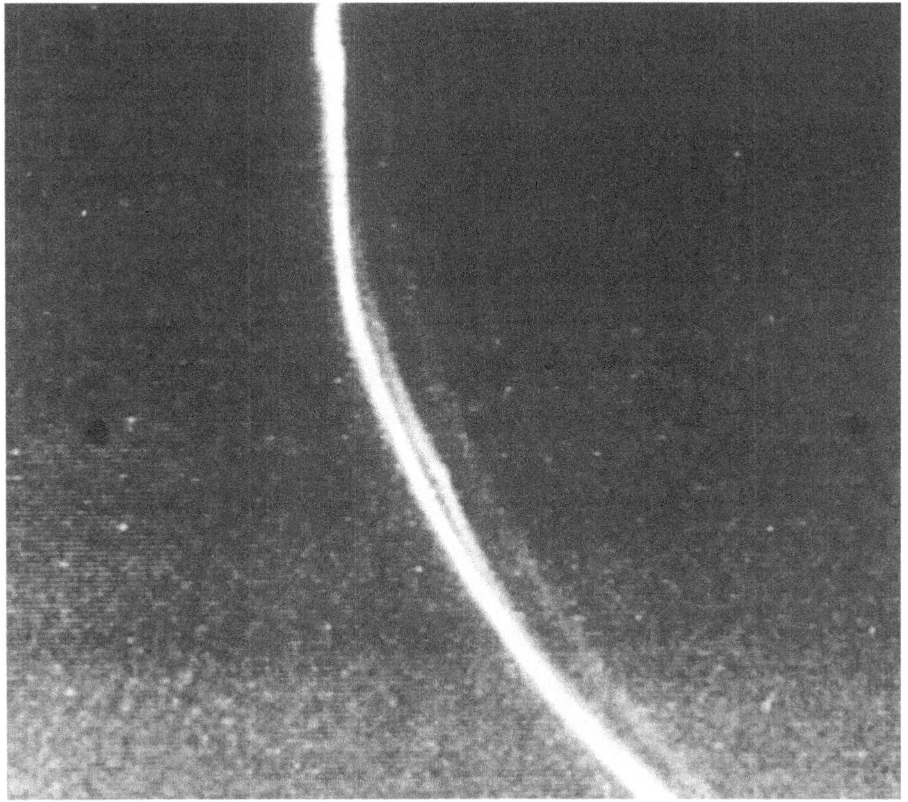

This Voyager 1 image clearly shows the kinky, braided structure of the F-ring.

Voyager's high-resolution azimuthal studies of the new F-ring showed that the ring was even more unusual than expected: It comprises a number of strands that appear to be intertwined and braided in some places, and kinked in other locations. While the gravitational effects of nearby moons may be responsible for some F-ring attributes, they do not account for the complex structure that Voyager discovered. Continuing studies of Voyager images over the years have unraveled more of the complex properties of this unusual ring.

The possibility of numerous, small satellites occurring within Saturn's ring system was a puzzle the Voyager mission had hoped to solve. Voyager's best-resolution studies of the ring system were aimed at revealing any bodies larger than about 10 kilometers in diameter; nevertheless, only four moonlets — Atlas, Pan, Prometheus and Pandora — were found in the images. Only one, Pan, was located in the main ring system.

The Voyager high-resolution studies did, however, detect signs of small moonlets not actually resolved in the images. When a small, dense body passes near a section of low-density ring material, its gravitational pull distorts the ring and creates what are known as "edge waves."[3] While the high-resolution, azimuthal coverage of the ring system is a time-consuming endeavor, such studies are the only means of detecting these waves and thereby finding satellites too small to be seen in images. Soon, four years in orbit will provide Cassini with many good opportunities for conducting azimuthal imaging studies.

Occultation Experiments

Cassini will be able to perform a number of the experiments that Voyager used to detect other gravitational effects on Saturn's ring material. One experiment, termed an "occultation experiment," involves watching as light from a star (a stellar occultation) or the radio beam from the spacecraft (a radio occultation) passes through the ring material to see how the beam is affected during the occultation. As the beam passes through the ring material, it may be attenuated or even extinguished. These experiments provided an extremely high-resolution study of a single ring path with resolutions up to about 100 meters.

SATURN'S CROWN JEWELS

Saturn's rings are easily the crown jewels of the entire solar system. And the variety of rings is staggering; the count in high-resolution images suggests anywhere from 500 to 1,000 separate rings! Named in order of discovery, the rings' labels do not indicate their relative positions. From the planet outward, they are known as D, C, B, A, F, G and E.

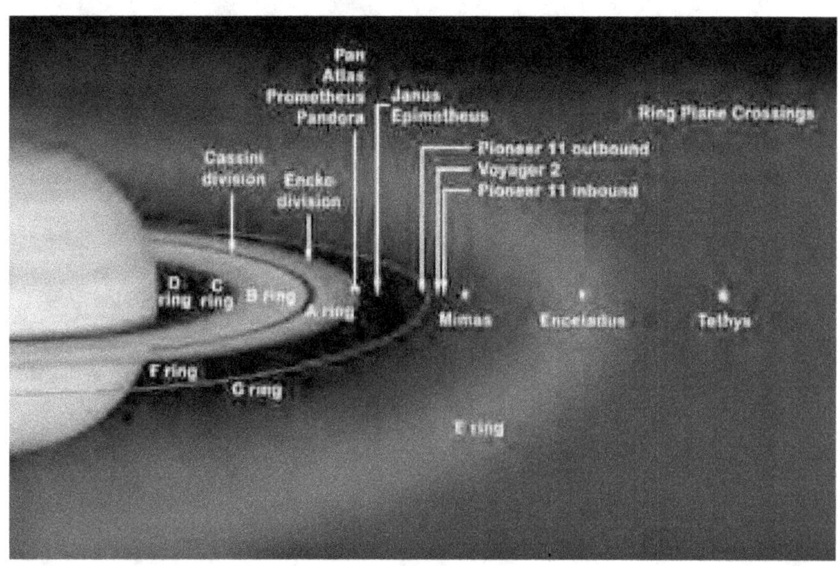

In the stellar occultation experiment, the Voyager 2 spacecraft had a single opportunity to observe as light from a star was occulted by the ring system. As the light passed through the rings, it was attenuated and extinguished by the particles comprising the system. The velocity of the spacecraft, the brightness of the star, the sensitivity of the instrumentation and the sampling rate of the instruments provide the limitations for such stellar occultation experiments.

The primary instruments used to conduct the stellar occultation study were Voyager's photopolarimeter and ultraviolet spectrometer, with the photopolarimeter providing the higher resolution data set. Resolving features as small as about 100 meters, the stellar occultation experiments detected many small-scale ring structures and found that the F-ring was far more complex than images had suggested. The data set, furthermore, showed that the B-ring was quite opaque in many regions and confirmed that the Cassini "division" was not at all empty. It also provided a direct measurement of the maximum thickness of the ring system in several locations, finding it to be much less than 100 meters thick.

In the radio occultation experiment, Voyager 1 had a single opportunity to watch as the spacecraft's radio beacon was occulted by the ring system. As this radio signal passed through the ring material, its signal was also attenuated. The antennas of the Deep Space Network received the attenuated signal, allowing scientists to study the complex data being returned. While the radio-science data set was complicated to process and interpret, it yielded information on particle size distribution in the ring system not available from stellar occultation studies. From the radio-science data set, we know something of the size distribution of the larger particles in the ring system. We know that much of the ring mass is in particles from a few centimeters to a few meters in diameter; boulders more than 10 meters in diameter do not comprise much of the mass of Saturn's ring system.

Both occultation experiments described here detected features in Saturn's rings caused by the gravitational effects of the planet's satellites. Locations where gravitational resonance effects had partially cleared material were identified by these experiments and are also visible at lower resolution in imaging data. While the existence of resonances had been recognized since 1866 — when it was pointed out that the Cassini division lay near a gravitational resonance point — the extent to which the structures occur was largely unexpected.

At these resonances, gravitational-wave effects were detected in the photopolarimeter, ultraviolet spectrometer and radio-science occultation data sets — known as density and bending waves — that are similar in behavior to the arms of spiral galaxies. While some of these structures were later detected in the imaging data, the high-resolution data sets of the photopolarimeter, ultraviolet spectrometer and radio-science experiments provided the details that allowed Voyager science teams to derive information about the mass, density and viscosity[4] of the ring material in their vicinity. Much of the fine structure of the A-ring is well-explained by such gravitational resonances with satellite orbits. However, the resonance theory does not provide a universal solution and much of the structure of the B-ring is left unexplained.

The occultation results, along with imaging studies, showed that a great many of the narrow ringlet features at Saturn are slightly eccentric — not circular in shape — and that these eccentric ringlets lie embedded in the mass of nearly circular rings that constitute the majority of Saturn's ring system. These same studies also showed that many small ringlets are asymmetrical in structure and that there are very few truly empty "gaps" in the ring system.

Results from the Voyager experiments provided a wealth of information to explain some physical properties of Saturn's rings, but at the same time, they also created many mysteries that were unexpected and not understandable through Voyager data. For instance, Voyager data did not detect the number of moonlets we expected. Are these moonlets nonexistent or just not yet discovered? Cassini will have far more time to perform ring searches and will provide higher resolution data sets, both required to answer this question.

Ring Spokes

As Voyager approached closer to Saturn, images began to show dark, radial structures on the rings. While these features had been sighted by Earth-based observers over the decades, the Voyager images showed the "spokes" in great detail, allowing them to be studied as they formed and rotated about the planet. Spokes seemed to appear rapidly — as a section of ring rotated out of the darkness near the dawn terminator — and then dissipate gradually, rotating around toward the dusk ring terminator. A spoke's formation time seemed to be very short; in some imaging studies one was seen to grow over 6000 kilometers in distance in just five minutes.

As it passed from the day side to the night side of the planet, Voyager provided unique views of Saturn's rings and was able to observe the rings and spokes in forward-scattered light where they appeared as far brighter features than in backscattered light. This light-scattering property showed that the spokes were largely composed of very small, micrometer-sized particles.

The numerous "spoke" features in the B-ring are evident in this image obtained by Voyager 2.

The spokes in Saturn's rings form at the distance from Saturn where the rotational speed of the ring particles matches that of the planet's magnetic field lines. Further studies of this possible interaction revealed a link between the spoke formation and a region on the planet associated with intense long-wavelength radio emissions (now referred to as the Saturn kilometric radiation, or SKR, zone). All these studies pointed to a sizable interaction between the ring system and Saturn's electromagnetic fields. But what was it? Once again, Voyager had provided a tantalizing tidbit of information, but had left a riddle.

Ring Particle Properties

The Voyager trajectories allowed the rings to be viewed from a much wider range of angles than could ever be achieved by Earth-based observers. This allowed for extensive observations of the light scattering, or photometric, properties of the rings. The proximity of the spacecraft to Saturn also provided for higher resolution photometric and spectral studies than could ever have been achieved by Earth-based observers. These studies yielded a great deal of information about the nature of the ring particles and the composition of the rings.

Observations in forward-scattered light made while the spacecraft was on the dark side of the planet showed that some areas of the rings were much brighter in forward-scattered light than others. Small particles in the micrometer-size range are much more efficient in producing forward scattering than are larger particles, so this tells us that these areas of the rings have an abundance of small, micrometer-size particles. The D, E, F and G rings all show such forward-scattering properties as do the spokes of the B-ring. These data, combined with ring occultation data from the photopolarimeter, ultraviolet spectrometer and radio-science experiments, give us some information about the variation of particle-size distribution across Saturn's ring system. Cassini will have the opportunity to conduct extensive studies of these diffuse rings and their particle distributions.

This false-color image of Saturn's B and C rings reveals fine details and subtle color differences.

Ring Composition

Ground-based infrared spectral studies of the A and B rings show that they are composed largely of very nearly pure water ice. The spectral characteristics of the rings are also very similar to those of several of Saturn's inner satellites. The extended wavelength range over which Cassini will be able to make spectral measurements will yield information on both the nature of the chemical contaminants in the rings' water ice as well as on the relationship between ring and satellite composition.

Studies of the color distribution (as a sign of compositional variation) in the main rings show that the ring system is not completely uniform in its makeup and that some sorting of materials within A and B rings exists. Why such a nonuniform composition exists is unknown; Cassini data will provide some clues toward solving this problem. The other, diffuse rings of Saturn are much more difficult to study in this manner, but we do know that the E-ring is somewhat bluish in color — and thus different in makeup from the main rings. Speculation exists that the moon Enceladus is the source of E-ring material, but this is unconfirmed.

Since ring particles larger than about one millimeter represent a considerable hazard to the Cassini spacecraft, the mission plan will include efforts to avoid dense particle areas of Saturn's ring plane. The spacecraft will be oriented so as to provide maximum protection for itself and its sensitive instrumentation packages. Even with such protective steps, passage through the ring plane will allow the Cassini particle measurement experiments (the Cosmic Dust Analyzer, the Ion and Neutral Mass Spectrometer and the Cassini Plasma Spectrometer) to perform important studies of the particles making up the less dense regions of the Cassini ring plane. Such measurements could provide insight into the composition and environment of the ring system and Saturn's icy satellites.

Voyager 1 looked back on November 16, 1980, four days after flying by the planet, to capture Saturn and its spectacular rings from this perspective.

Formation and Evolution

A century after their discovery, the origin of Saturn's rings was a point of speculation. Then, for many years, the matter was actually thought to be well understood. Today's view is that the origin of those magnificent rings involves a complex set of issues. The size, mass and composition of the rings make their formation and evolution rather difficult to explain.

Among the early notions about the rings' origin was a theory by Edouard Roche. He suggested that the rings were fragments left over from a moon that had at one time orbited Saturn. This moon had broken up, Roche explained, because of the tremendous stresses placed upon it by Saturn's huge gravitational field.

The Roche theory does not hold up well, however, under scrutiny. For instance, did the rings form out of the initial solar system nebula, or after one or more satellites were torn apart by Saturn's gravity? Moreover, if the rings were the result of the numerous comets captured and destroyed by Saturn's gravity, why are Saturn's rings so different in nature from the rings of the other giant planets? Over their lifetime, the rings must have been bombarded continually by comets and meteors — they should have accumulated a great amount of carbonaceous and silicate debris — yet their composition seems almost entirely to be water ice.

Another issue concerns the stability of the ring system. The effects of torque and gravitational drag, along with the loss of momentum via collisional processes should have produced a system only one-tenth to one-hundredth the age of the solar system itself. If this hypothesis is correct, then we cannot now be observing a ring system around Saturn that formed when the solar system coalesced.

In fact, Saturn's rings — as well as the rings of all the other large planets — may have formed and dissipated many times since the beginning of the solar system. If anything, we may infer that ring systems are in a constant, steady state of renewal and regeneration. We are quite likely a long way from understanding all the processes and dynamics of ring formation, renewal and evolution.

In conducting an array of ring studies, the Cassini mission will ultimately focus on four critical questions: How did the rings form? How old are the rings? How are the rings maintained? What are the dynamics and relationships of the rings to Saturn, its satellites and its electromagnetic fields?

In attempting to answer these questions, whereas the Voyager spacecraft had only a few days close to Saturn's rings — time only enough to suggest some tantalizing clues — Cassini will have several months to observe the rings in detail and return enough data to substantially increase our knowledge base.

Voyager Legacy, Cassini Mission

Information about Saturn's rings "discovered" by the Voyager mission — which was then studied and documented — was not entirely new to Earth-bound astronomers. Ring gaps, narrow rings, spokes, diffuse rings, sharp ring edges, satellite resonances — these were all features seen or inferred before. Nevertheless, scientists were stunned by the sheer abundance of these features on every scale once Voyager began returning actual, high-resolution images of the ring system.

As more and more data arrived on Earth, one thing was increasingly clear: Rather than solving the mystery of Saturn's rings, Voyager was instead showing us how little we knew about them, even after four centuries of study!

After two decades of studying the Voyager data, and then studying more recent data from the Hubble Space Telescope, scientists have developed a new set of questions for the Cassini mission to investigate when the spacecraft reaches Saturn in 2004. The answers Cassini may find to these questions will help to explain some of the processes in the rings' origin. Scientists are aware, however, that the data Cassini returns will quite probably present us with another round of surprises — and yet more questions.

Early observers were puzzled by the changes that Saturn's rings showed

VOYAGER–THE GRANDEST TOUR

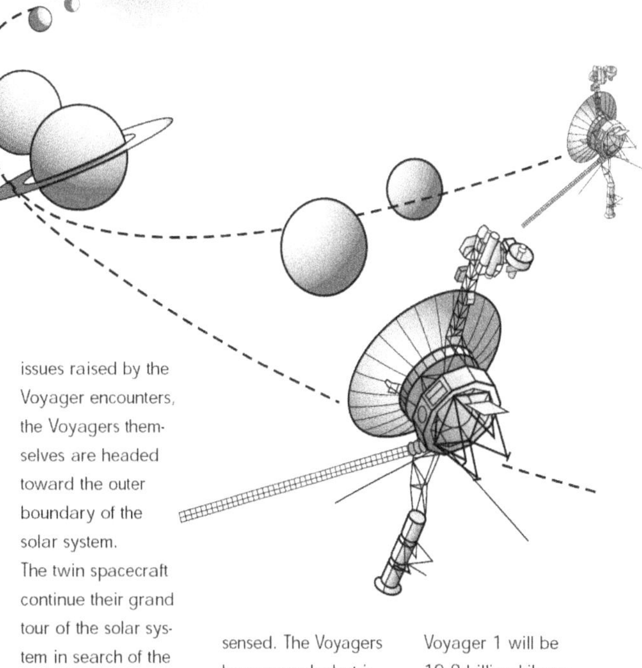

Scientists have long sought to determine what the rings of Saturn are made of, how they got there and what keeps them in orbit around the giant planet. Some of the questions began to be answered when Voyager 1 arrived at Saturn on November 12, 1980, and Voyager 2 followed suit on August 25, 1981. The Voyagers' images revealed the rings as the most exquisite sight in the solar system. Still, the discoveries presented many more new puzzles than solutions.

Now, as Cassini–Huygens prepares to tackle the many issues raised by the Voyager encounters, the Voyagers themselves are headed toward the outer boundary of the solar system. The twin spacecraft continue their grand tour of the solar system in search of the heliopause, the region where the Sun's influence wanes and the beginning of interstellar space can be sensed. The Voyagers have enough electrical power and thruster fuel to operate until at least 2015. By then, Voyager 1 will be 19.8 billion kilometers away from the Sun, and Voyager 2 will be 16.8 billion kilometers away.

from year to year. Now, after centuries of Earth-bound observations followed by spacecraft explorations, scientists do not question if Cassini will find changes in Saturn's ring system. Instead, they wonder what changes the rings have undergone since Voyager last looked in 1981.

Among other things, Cassini will probe the many questions raised by Voyager, and Pioneer before that, about Saturn's rings. In this endeavor, the new spacecraft has several expected advantages over its predecessors. Cassini's science instrumentation covers a broader range of the electromagnetic spectrum and is of greater sensitivity and resolution.

In addition, while Voyager was a mission of discovery into previously uncharted frontiers, Cassini is a focused return visit, designed to address specific issues. Of course, Cassini is also expected to make its own discoveries about Saturn.

The ring system of Saturn — which contains the widest array of attributes of all the ring systems known to us — is the ideal laboratory for studying the ring phenomenon. The Voyager mission showed us how little we really know about Saturn's rings, even after four centuries of observations. While Cassini provides us with a wealth of information to help solve some of the rings' mystery, this new mission will also uncover many new clues for future investigation.

For now, scientists anxiously await the volumes of data — about those magnificent rings — that Cassini is expected to return to Earth between 2004 and 2008.

FOOTNOTES

[1] If a satellite and a ring particle have a rotational period that are integer multiples of one another, their periods are said to be in resonance.

If a ring particle orbits in exactly one-half the amount of time it takes a given satellite to complete its orbit, we have a 2:1 resonance. The ring particles in such an orbit obtain repeated gravitational "tugs," which create more collisions, which eventually thin the region of particles. The Cassini gap is the best known and best defined such zone, with the Janus 7:6 point at its exterior limit and the Mimas 2:1 point at its interior limit. In regions where the resonance effect is less pronounced, we find density waves instead of gaps in the rings.

[2] Keeler probably discovered what is known as the Encke gap, but the astronomical community has been somewhat remiss in correcting this error.

[3] Further searching in the Encke gap, where such edge waves were seen, led to the discovery of one small satellite inside the division. Called Pan, this satellite may well be responsible for clearing and maintaining the gap.

[4] The particles constituting the rings are not completely isolated from each other; they interact with one another in orbit through collisions and their mutual gravitational attraction. These interactions make the rings behave somewhat like a fluid, rather than a swarm of isolated, orbiting particles. The degree to which the particles interact is characterized as a measure of the rings' "viscosity."

CHAPTER 5

The Icy Satellites

The moons, or satellites, of Saturn represent a diverse set of objects. They range from the planet-like Titan, which has a dense atmosphere, to tiny, irregular objects only tens of kilometers in diameter. These bodies are all believed to have as major components some type of frozen volatile, primarily water ice, but also other components such as methane, ammonia and carbon dioxide existing either alone or in combination with other volatiles. Saturn has at least 18 satellites; there may exist other small, undiscovered satellites in the planet's system.

Overall Characteristics

In 1655, Christiaan Huygens discovered Titan, the giant satellite of Saturn. Later in the 17th century, Jean-Dominique Cassini discovered the four next largest satellites of Saturn. It was not until more than 100 years later that two smaller moons of Saturn were discovered. As telescopes acquired more resolving power in the 19th century, the family of Saturn's satellites grew. Most of the smallest satellites were discovered during the Voyager spacecraft flybys. The 18th satellite, Pan, was found nearly 10 years after the flybys during close analysis of Voyager images; it is embedded in the A-ring of Saturn. Saturn's ring plane crossings — when the obscuring light from Saturn's bright rings dims, as the rings move to an edge-on orientation — represent the ideal configuration for discovering new satellites. Images obtained by the Hubble Space Telescope during the ring plane crossings in 1995 did not reveal any unambiguous discoveries of new satellites.

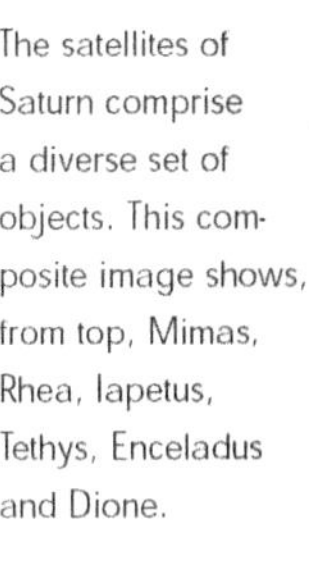

The satellites of Saturn comprise a diverse set of objects. This composite image shows, from top, Mimas, Rhea, Iapetus, Tethys, Enceladus and Dione.

Physical and Dynamic Properties

Most planetary satellites present the same hemisphere toward their primaries, a configuration that is the result of tidal evolution. When two celestial bodies orbit each other, the gravitational force exerted on the near side is greater than that exerted on the far side. The result is an elongation of each body that forms tidal bulges, which can consist of either solid, liquid or gaseous (atmospheric) material. The primary tugs on the satellite's tidal bulge to lock its longest axis on to the primary satellite line. The satellite, which is said to be in a state of synchronous rotation, keeps the same face toward the primary. Since this despun state occurs rapidly (usually within a few million years), most natural satellites are in synchronous rotation. Of Saturn's icy satellites, two, Hyperion and Phoebe, are known to exhibit asynchronous rotation.

The Hubble Space Telescope obtained these images of Saturn and its rings and satellites. The top image, obtained in August 1995, shows Titan (far left, casting a shadow) and (far right, from the left), Mimas, Tethys, Janus and Enceladus. In the middle image, obtained in November 1995, Tethys (upper left) and Dione appear. In the bottom image, obtained in December 1994, a storm is raging in the center of the planet. [Images courtesy of: top and middle — E. Karkoschka, University of Arizona Lunar and Planetary Laboratory; bottom — R. Beebe, New Mexico State University]

Before the advent of spacecraft exploration, planetary scientists expected the satellites of the outer planets to be geologically dead. They assumed that heat sources were not sufficient to have melted the satellites' mantles enough to provide a source of liquid or even semiliquid ice or ice silicate slurries. The Voyager and Galileo spacecraft have radically altered this view by revealing a wide range of geological processes on the moons of the outer planets. Enceladus may be currently active. Several of Saturn's medium-sized satellites are large enough to have undergone internal melting with subsequent differentiation and resurfacing.

Recent work on the importance of tidal interactions and subsequent heating has provided the theoretical foundation for explaining the existence of widespread activity in the outer solar system. Another factor is the presence of non-ice components, such as ammonia hydrate or methane clathrate, which lower the melting points of near-surface materials. Partial melts of water ice and various contaminants — each with their own melting point and viscosity — provide material for a wide range of geological activity.

Hyperion is an irregular, pockmarked satellite in an unusual, nonsynchronous rotational state.

Because the surfaces of so many outer planet satellites exhibit evidence of geological activity, planetary scientists have begun to think in terms of unified geological processes occurring on the planets — including Earth — and their satellites. For example, partial melts of water ice with various contaminants could provide flows of liquid or partially molten slurries that in many ways mimic the terrestrial or lunar lava flows formed by the partial melting of silicate rock mixtures. The ridged and grooved terrains on satellites such as Enceladus and Tethys may have resulted from tectonic activities occurring through-

THE KNOWN MOONS OF SATURN

Moon	Diameter, kilometers	Distance, kilometers	Orbital Period, days	Year Discovered: Discoverer
Pan	20	133,580	0.56	1990: Showalter
Atlas	30	137,670	0.60	1980: Terrile
Prometheus	100	139,350	0.61	1980: Collins
Pandora	90	141,700	0.63	1980: Collins
Epimetheus	120	151,450	0.69	1966: Walker
Janus	190	151,450	0.69	1966: Dolfus
Mimas	392	185,520	0.94	1789: Herschel
Enceladus	500	238,020	1.37	1789: Herschel
Tethys	1060	294,660	1.89	1684: Cassini
Telesto	30	294,660	1.89	1980: Smith
Calypso	26	294,660	1.89	1980: Smith
Dione	1120	377,400	2.74	1684: Cassini
Helene	32	377,400	2.74	1980: Laques and Lecacheux
Rhea	1530	527,040	4.52	1672: Cassini
Titan	5150	1,221,830	15.94	1655: Huygens
Hyperion	290	1,481,100	21.28	1848: Bond
Iapetus	1460	3,561,300	79.33	1671: Cassini
Phoebe	220	12,952,000	550.40	1898: Pickering

out the solar system. Finally, the explosive volcanic eruptions possibly occurring on Enceladus may be similar to those occurring on Earth, which result from the escape of volatiles released as the pressure decreases in upward-moving liquids.

Formation and Bulk Composition

The solar system — the Sun, the planets and their families of moons — condensed from a cloud of gas and dust about 4.6 billion years ago. This age is derived primarily from radiometric dating of meteorites, which are believed to consist of primordial, unaltered matter. Because there existed a temperature gradient in the protosolar cloud, or nebula, volatile materials (those with low condensation temperatures) are the major components of bodies in the outer solar system. These materials include water ice, ice silicate mixtures, methane, ammonia and the hydrated forms of the latter two materials. Two types of dark contaminants exist on the surfaces of the satellites: C-type or carbon-rich material, and D-type material, which is believed to be rich in hydrocarbons.

The planets and their satellite systems formed from the accretion of successively larger blocks of material, or planetesimals. One important concept of planetary satellite formation is that a satellite cannot accrete within Roche's limit, the distance at which the tidal forces of the primary become greater than the internal cohesive forces of the satellite. Except for Titan, Saturn's satellites are too small to possess gravity sufficiently strong to retain an appreciable atmosphere against thermal escape.

Evolution

Soon after the satellites accreted, they began to heat up from the release of gravitational potential energy. An additional heat source was provided by the release of mechanical energy during the heavy bombardment of their surfaces by remaining debris. Mimas and Tethys have impact craters (named Herschel and Odysseus, respectively) caused by bodies that were nearly large enough to break them apart; probably such catastrophes did occur. The decay of radioactive elements found in silicate materials provided another major source of heat. The heat produced in the larger satellites may have been sufficient to cause melting and chemical fractionation; the dense material, such as silicates and iron, went to the center of the satellite to form a core, while ice and other volatiles remained in the crust. A fourth source of heat is provided by the release of frictional energy as heat during tidal and resonant interactions among the satellites and Saturn.

Several of Saturn's satellites underwent periods of melting and active geology within a billion years of their formation and then became quiescent. Enceladus may be currently geologically active. For nearly a billion years after their formation, the satellites all underwent intense bombardment and cratering. The bombardment tapered off to a slower rate and presently continues. By counting the number of craters on a satellite's surface and making certain assumptions about the flux of impacting ma-

SMALL ONES, BIG ONES...

The moons, or satellites, of Saturn comprise a diverse set of objects. Of the 18 currently known satellites, nine are illustrated below to show their relative sizes. The planet-like Titan is easily the largest of Saturn's moons. It has a dense atmosphere and possibly liquid oceans on its surface. Hyperion, by contrast, is irregularly shaped and seems to be covered with ice and dark, rocky material. The Cassini–Huygens mission, through its detailed observations, will provide us with much more information about Saturn's many moons.

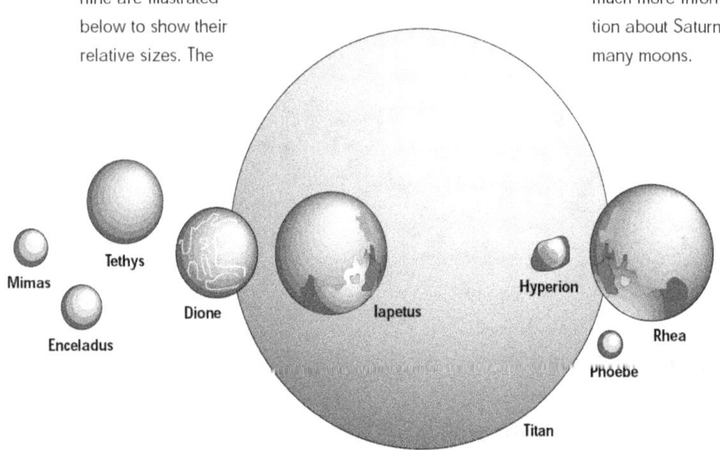

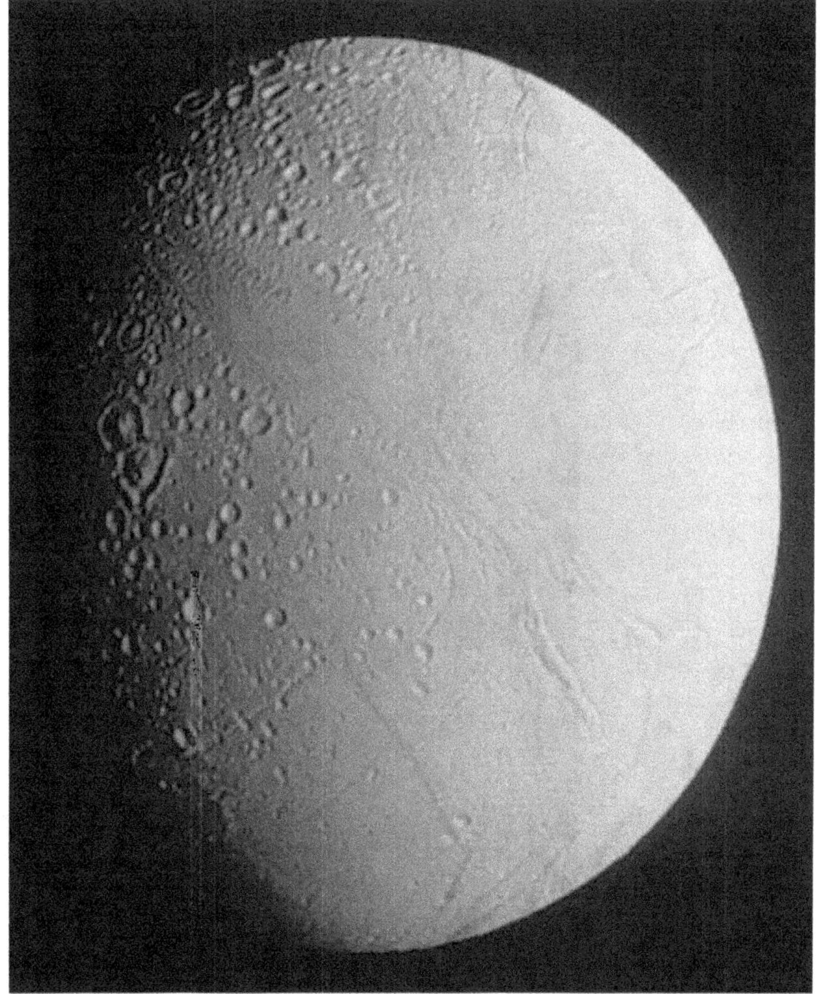

The bright surface of Saturn's moon Enceladus shows evidence of extensive resurfacing over time.

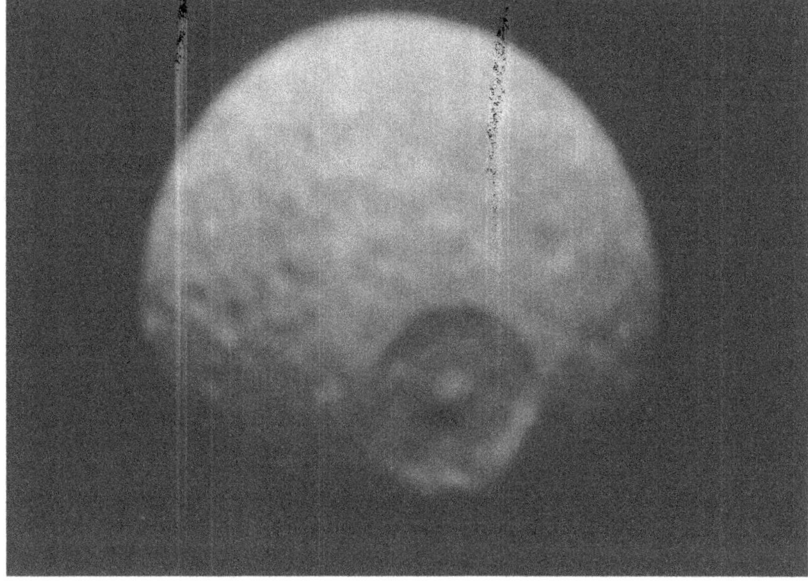

The Herschel crater covers approximately one third of the diameter of the surface on the moon Mimas.

terial, geologists are able to estimate when a specific portion of a satellite's surface was formed. Continual bombardment of satellites causes the pulverization of both rocky and icy surfaces to form a covering of fine material known as a regolith.

Meteoritic bombardment of icy bodies alters the optical characteristics of the surface by excavating and exposing fresh material. Impacts can also cause volatilization and the subsequent escape of volatiles to create a lag deposit enriched in opaque, dark materials. Both the Galilean satellites of Jupiter and the medium-sized satellites of Saturn tend to be brighter on the hemispheres leading in the direction of orbital motion (the so-called "leading" side, as opposed to the "trailing" side); this effect is thought to be due to preferential micrometeoritic "gardening" on the leading side.

Many geologists expected the craters formed on the outer planets' satellites to have disappeared from viscous relaxation. Voyager images reveal craters that in many cases have morphological similarities to those found in the inner solar system, including central pits, large ejecta blankets and well-formed outer walls. Scientists now believe that silicate mineral contaminants or other impurities in the ice provide the extra strength required to sustain impact structures.

Individual Satellites

Medium-Sized Icy Satellites. These six satellites of Saturn are smaller than Titan and Jupiter's giant Galilean satellites, but they are still sizable — and as such they represent a unique class of icy satellite. Earth-based telescopic measurements showed the spectral signature of ice for Tethys, Rhea and Iapetus; Mimas and Enceladus are close to Saturn and difficult to observe because of scattered light from the planet. The satellites' low densities and high albedos imply that their bulk composition is largely water ice, possibly combined with ammonia or other volatiles. They have smaller amounts of rocky silicates than the

The six medium-sized icy satellites of Saturn. Clockwise from upper left: Rhea, Tethys, Mimas, Enceladus, Dione and Iapetus.

Artist's view of Saturn as seen from the planet's heavily cratered moon, Mimas. In the middle of the image is a central pit of a large impact crater.

Galilean satellites. Most of what is presently known of the Saturn system was obtained from the Voyager flybys in 1980 and 1981.

Saturn's innermost, medium-sized satellite, Mimas, is covered with craters, including one named Herschel that covers a third of the moon's diameter. There is a suggestion of surficial grooves that may be features caused by the impact. The craters on Mimas tend to be high-rimmed, bowl-shaped pits; apparently surface gravity is not sufficient to have caused viscous relaxation. The application of crater-counting techniques to Mimas suggests that it has undergone several episodes of resurfacing.

The next satellite outward from Saturn is Enceladus, an object that was known from telescopic measurements to reflect nearly 100 percent of the visible radiation incident on it (for comparison, Earth's Moon reflects only about 11 percent). The only likely composition consistent with this observation is almost pure water ice or some other highly reflective volatile. Voyager 2 images show an object that had been subjected, in the recent geological past, to extensive resurfacing; grooved formations similar to those on the Galilean satellite Ganymede are prominent.

The lack of impact craters on this terrain is consistent with an age of less than a billion years. Some form of ice volcanism may be currently occurring on Enceladus. A possible heating mechanism is tidal interactions, perhaps with Dione. About half of the surface observed by Voyager is extensively cratered, consistent with an age of four billion years.

A final element to the puzzle of Enceladus is the possibility that it is responsible for the formation of the E-ring of Saturn, a tenuous collection of icy particles that extends from inside the orbit of Enceladus to past the orbit of Dione. The maximum thickness position of the ring coincides with the orbital position of Enceladus. If some form of volcanism is presently active on the surface, it could provide a source of particles for the ring. An alternative source mechanism is the escape of particles from the surface due to meteoritic impacts.

Tethys is covered with impact craters, including Odysseus, the largest known impact structure in the solar system. The craters tend to be flatter than those on Mimas or the Moon, probably because of viscous relaxation and flow over the eons under Tethys' stronger gravitational field. Evidence for episodes of resurfacing is seen in regions that have fewer craters and higher albedos. In addition, there is a huge trench formation, the Ithaca Chasma, which may be a degraded form of the grooves found on Enceladus.

Dione, about the same size as Tethys, but more dense, exhibits a wide diversity of surface morphology. Next to Enceladus, it has the most extensive evidence for internal activity. Its relatively high density may provide added radiogenic heat from siliceous material to spur this activity. Most of the surface is heavily cratered, but gradations in crater density indicate that several periods of resurfacing occurred during the first billion years of its existence. The leading side of the satellite is about 25 percent brighter than the other, due possibly to more intensive micrometeoritic bombardment on this hemisphere. Wispy streaks, which are about 50 percent brighter than the surrounding areas, are believed to be the result of internal activity and subsequent emplacement of erupting material. Dione modulates the radio emission from Saturn, but the mechanism responsible for this phenomenon is unknown.

Rhea appears to be superficially very similar to Dione. Bright wispy streaks cover one hemisphere. However, there is no evidence for any resurfacing events early in its history. There does seem to be a dichotomy between crater sizes — some regions lack large craters while other regions have a preponderance of such impacts. The larger craters may be due to a population of larger debris more prevalent during an earlier episode of collisions. The craters on Rhea show no signs of viscous relaxation.

When Cassini discovered Iapetus in 1672, he noticed almost immediately that at one point in its orbit around Saturn it was very bright, but on the opposite side of the orbit, the moon nearly disappeared. He correctly de-

The cratered surface of Tethys, including the groove-like Ithaca Chasma and the crater Telemachus at the upper right.

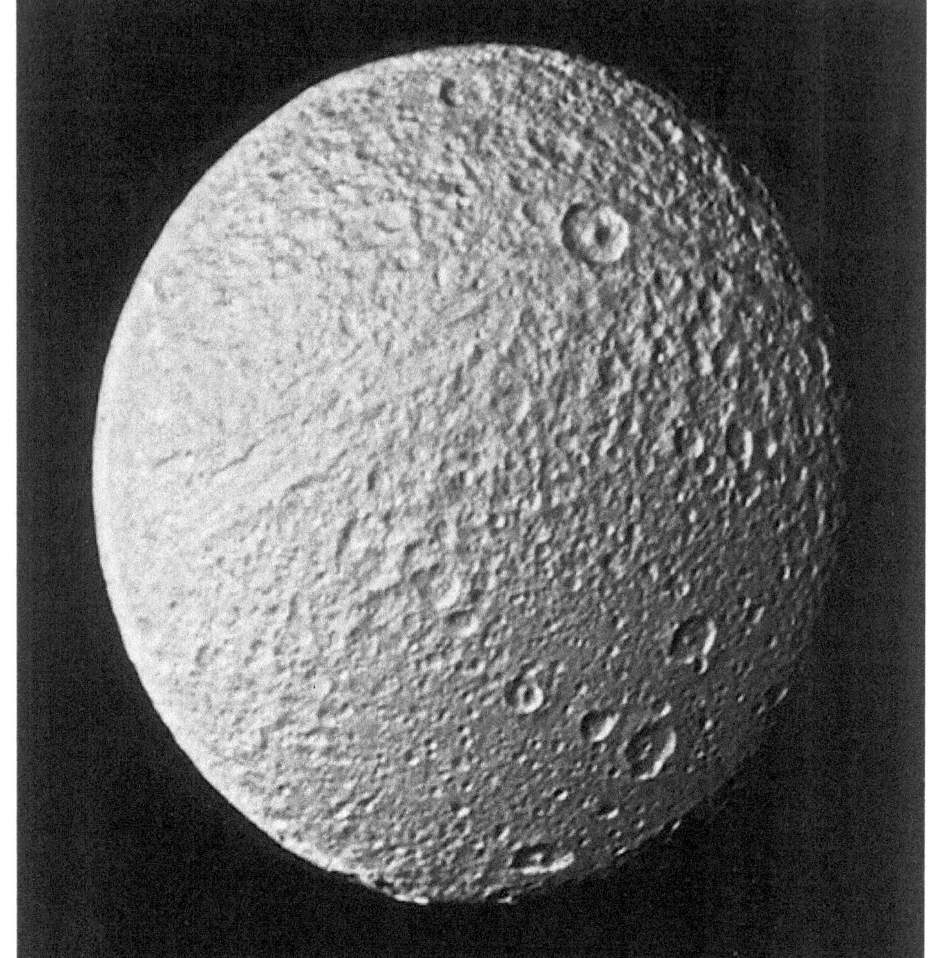

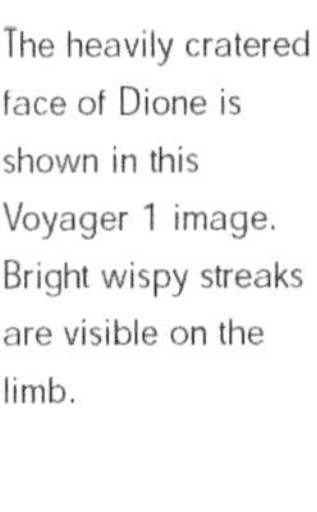

The heavily cratered face of Dione is shown in this Voyager 1 image. Bright wispy streaks are visible on the limb.

duced that one hemisphere is composed of highly reflective material, while the other side is much darker. Voyager images show that the bright side, which reflects nearly 50 percent of the incident radiation, is fairly typical of a heavily cratered icy satellite. The other side, which is centered on the direction of motion, is coated with a redder material that has a reflectivity of about three to four percent.

Scientists do not agree on whether the dark material originated from an exogenic source or was endogenically created. One scenario for the exogenic deposit of material entails dark particles being ejected from Phoebe and drifting inward to coat Iapetus. The major problem with this model is that the dark material on Iapetus is redder than Phoebe, although the material could have undergone chemical changes after its expulsion from Phoebe that made it redder. One observation lending credence to an internal origin is the concentration of material on crater floors, which implies an infilling mechanism. In one model, methane erupts from the interior and is subsequently darkened by ultraviolet radiation.

Other characteristics of Iapetus are odd. It is the only large Saturn satellite in a highly inclined orbit. It is less dense than objects of similar albedo; this fact implies a higher fraction of ice or possibly methane or ammonia in its interior.

Small Satellites. The Saturn system has a number of unique small satellites. Three types of objects have been found only in the Saturn system: the shepherding satellites, the co-orbitals and the Lagrangians. All three groups of satellites are irregularly shaped and probably consist primarily of ice.

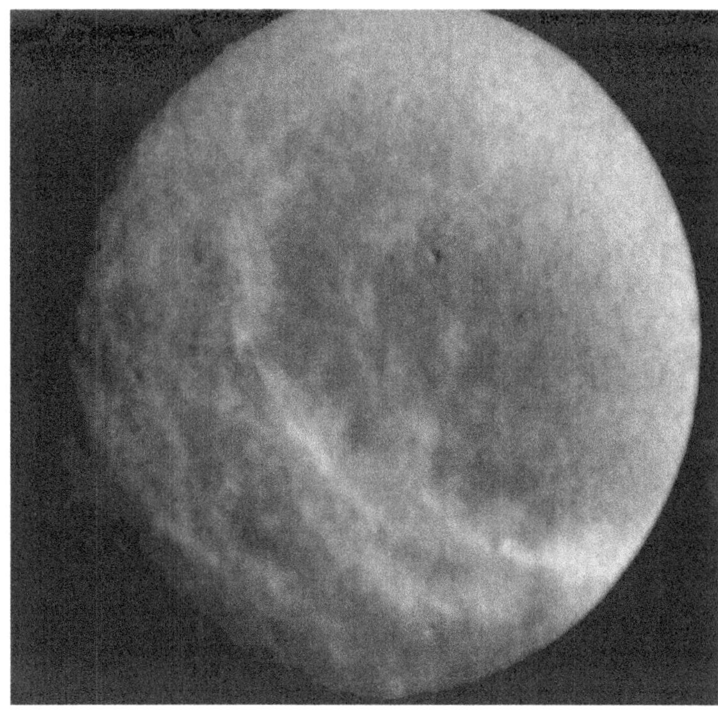

An array of bright streaks is visible in this view of Rhea, Saturn's second largest satellite.

The three shepherds, Atlas, Pandora and Prometheus, are believed to play a key role in defining the edges of Saturn's A and F rings. The orbit of Saturn's second innermost satellite, Atlas, lies several hundred kilometers from the outer edge of the A-ring. The other two shepherds, which orbit on either side of the F-ring, constrain the width of this narrow ring and may cause its kinky appearance.

The co-orbital satellites, Janus and Epimetheus, which were discovered in 1966 and 1978, respectively, exist in an unusual dynamic situation. They move in almost identical orbits at about two and a half Saturn radii. Every four years, the inner satellite (which orbits slightly faster than the outer one) overtakes its companion. Instead of colliding, the satellites exchange orbits. The four-year cycle then begins again. Perhaps these two satellites were once part of a larger body that disintegrated after a major collision.

The three other small satellites of Saturn — Calypso, Helene and Telesto — orbit in the Lagrangian points of larger satellites, one associated with Dione and two with Tethys. Lagrangian points are locations within an object's orbit in which a less massive body can move in an identical, stable orbit. The points lie about 60 degrees in front of and behind the larger body. Although no other known satellites in the solar system are Lagrangians, the Trojan asteroids orbit in two of the Lagrangian points of Jupiter.

Telescope observations showed that the surface of Hyperion, which lies between the orbits of Iapetus and Titan, is covered with ice. Because Hyperion has a relatively low albedo, however, this ice must be mixed with a significant amount of darker, rocky material. The color of Hyperion is similar to that of the dark side of Iapetus and D-type asteroids: All three bodies may be rich in primitive material rich in organics. It is darker than the medium-sized, inner satellites, presumably because resurfacing events have never covered it with fresh ice.

Although Hyperion is only slightly smaller than Mimas, it has a highly irregular shape, which along with the satellite's battered appearance, suggests that it has been subjected to intense bombardment and fragmentation. There is also good evidence that Hyperion is in a chaotic rotation, perhaps a collision within the last few million years knocked it out of a tidally locked orbit.

Saturn's outermost satellite, Phoebe, a dark object with a surface composition probably similar to that of C-type asteroids, moves in a highly inclined, retrograde orbit, suggesting it is a captured object. Voyager images show definite variegations consisting of dark and bright (presumably icy)

HOW TO NAME A MOON

Planetary satellites, including Saturn's, are generally named after figures in classical Greek or Roman mythology associated with the namesakes of their primaries. Hyperion, Iapetus, Phoebe, Rhea and Tethys, for example, were siblings of Kronos, the Greek counterpart of the Roman god Saturn.

The satellites also carry scientific designations, which comprise the first letter of the primary followed by a sequential Arabic numeral, assigned in order of discovery: Mimas is S1, Titan is S2 and so on. When satellites are first discovered, but not yet confirmed or officially named, they are known by the year in which they were discovered, the initial of the primary and a number assigned consecutively for all solar system discoveries. So, Pan was first called 1981S13 (although Pan was discovered in 1990, the Voyager image in which the moon was found was obtained in 1981).

After planetary scientists were able to map geological formations of the satellites from spacecraft images, they named many of the features after characters or locations from world mythologies. The official names for all satellites and surface features are assigned by the International Astronomical Union.

patches on the surface. Although it is smaller than Hyperion, Phoebe has a nearly spherical shape. With a rotation period of about nine hours, Phoebe is the only Saturn satellite known to exhibit a simple, asynchronous rotation.

Pan, the 18th known satellite of Saturn, was discovered in 1990 in Voyager 2 images that were obtained in 1981. This small object is embedded within the A-ring and helps to clear the Encke division of particles.

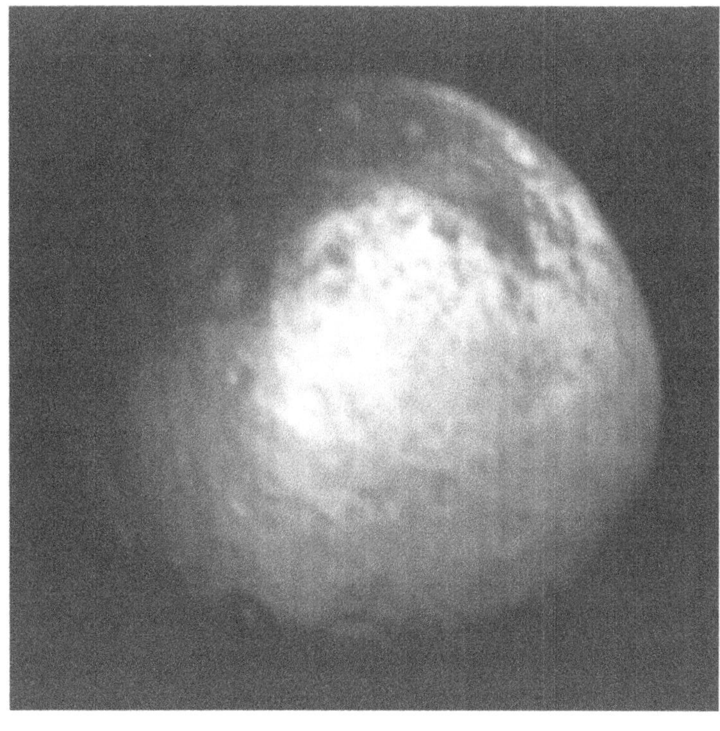

This Voyager 2 image of Iapetus shows both bright and dark terrains on the moon.

The Cassini–Huygens Mission

The two Voyager spacecraft provided the first detailed reconnaissance of the Saturn system. They provided the first evidence that the satellites had been geologically active after their formation and that one icy satellite, Enceladus, may be still active. There are significant gaps in our knowledge about Saturn's moons that can only be addressed by a more detailed observational plan and sophisticated instruments. The Cassini mission is designed to undertake this endeavor.

The Cassini payload represents a carefully chosen suite of instruments that will address the major scientific questions surrounding Saturn's moons, as follows:

First, what is the composition of the surfaces of these satellites? Although ground-based spectroscopic measurements showed water ice to be prevalent on their surfaces, the existence of additional volatiles, hydrates, clathrates and impurities is a critical factor in understanding the satellites' evolution. Because many impurities lower the freezing point of ice, their existence could provide an explanation for the satellites' activity. Hydrated ammonia, for example, could be the driver for activity on Enceladus and Dione, but its spectroscopic signature is so subtly different from water ice, that only close reconnaissance by Cassini's Visible and Infrared Mapping Spectrometer (VIMS) may be able to detect its presence.

In most cases, the satellites are too dim for detailed Earth-based spectroscopic studies. Much of the key spectroscopic evidence exists in the ultraviolet and 3–20-micrometer region, which is opaque to the terrestrial atmosphere. The nature of the dark material on the dark side of Iapetus, Hyperion, Phoebe and some of the small satellites is mysterious. Is it unprocessed and primordial? Is this material rich in organics and, as such, is it related to the origin of life? Is it similar to the dark material found on comets, some asteroids and other satellites in the outer solar system?

Next, what is the detailed morphology of the satellites, and what is the relationship between geological structures and compositional units? With a synergistic payload offering an Imaging Science Subsystem (ISS) that is capable of tens of meters resolution on the satellites and high-resolution spectrometers in the 0.055–1000 micrometer region — the Composite Infrared Spectrometer (CIRS), the Ultraviolet Imaging Spectrograph (UVIS) and the VIMS — and complementary radio-metric radar K_u-band measurements, it will be possible to correlate geological units with specific compositions.

The small satellites of Saturn. Clockwise from left: Atlas, Pandora, Janus, Calypso, Helene, Telesto, Epimetheus and Prometheus.

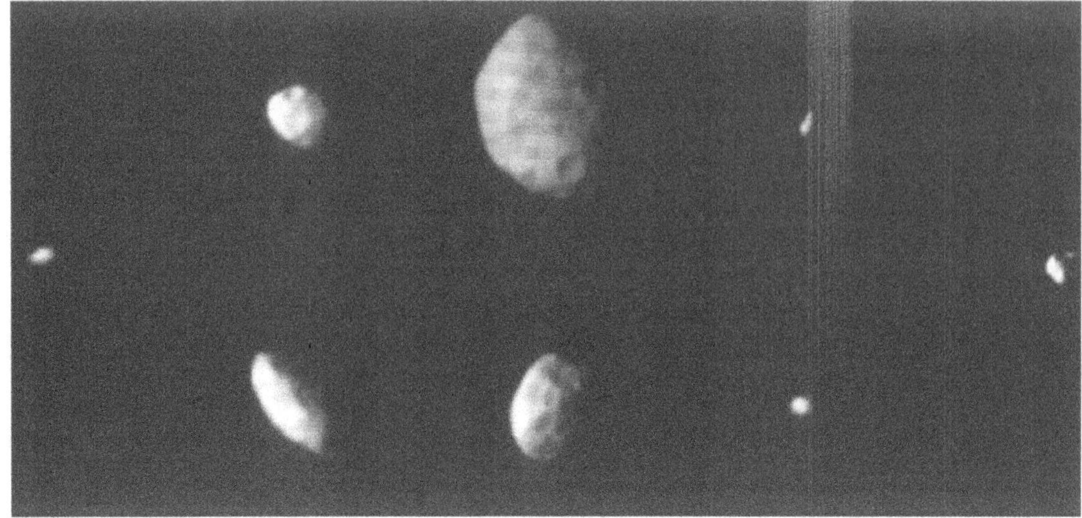

For example, what is the detailed distribution of dark material on Iapetus' craters, and what does that say about the origin of the material? Are the bright wispy streaks of Dione and Rhea rich in ammonia and thus plausibly formed from a hydrated ammonia slurry exuded onto the surface? What is the distribution of surficial impurities, and can they offer an explanation for the varying degrees of viscous relaxation on the satellites? Are the inner satellites coated with material from the E-ring?

Third, are there compositional similarities between the rings, dust and some of the satellites, and if so, does that imply they are interrelated? Are the rings comminuted satellites? Do any of the dust or icy particles, which will be measured in detail by Cassini's Cosmic Dust Analyzer (CDA), appear to come from satellites? Is Enceladus the source of the E-ring?

Also, do any of the satellites have thin, tenuous atmospheres of molecular oxygen, OH or other material, similar to those recently found on Jupiter's moons Europa and Ganymede? Cassini's UVIS or possibly the ISS should be able to detect such atmospheres, which may in turn provide material to Saturn's magnetosphere. If there is significant erosion of material from the satellites' surfaces (or in the case of Enceladus, expulsion), the Ion and Neutral Mass Spectrometer (INMS), the CDA and the Radio and Plasma Wave Science instrument (RPWS) will study their composition, size and mass.

Fifth, are there any additional satellites? Are there any embedded in the ring system? How do these satellites interact dynamically with the rings? High-resolution imaging by the ISS will be able to detect kilometer-sized bodies and resolve the issue of 10 or so unconfirmed observations of additional satellites.

Further, what is the relationship between Saturn's magnetosphere and the satellites? Are the satellites a significant source of magnetospheric particles? Do any of the satellites have magnetic fields? Why does Dione modulate Saturn's radio emission? Is there a flux tube of some sort between the satellite and the planetary magnetic field? The Dual Technique Magnetometer (MAG), the Magnetospheric Imaging Instrument (MIMI), the Cassini Plasma Spectrometer (CAPS), the Ion and Neutral Mass Spectrometer (INMS) and the RPWS will be the key instruments for answering these questions.

Seventh, is Enceladus still active? High-resolution imaging by the ISS will detect any recent features, as well as geysers or plumes. Further compositional analysis of the plumes and deposits will help to determine the physical mechanism of the activity and its similarity to other activity on Jupiter's moon Io, Neptune's moon Triton and Earth.

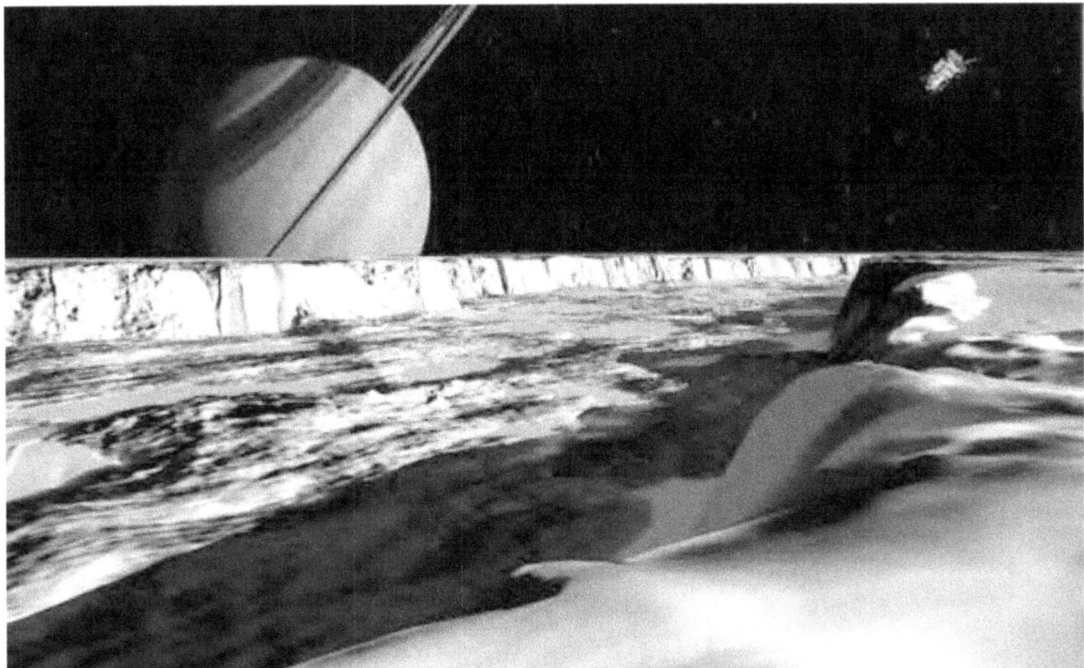

Artist's conception of Ithaca Chasma, a huge trench on the moon Tethys.

This artist's rendition shows Saturn as it might appear from the surface of its second largest satellite, Rhea.

Then, what are the internal structures of the satellites? Did they fully differentiate? Close flybys of the satellites entailing precise measurements by the RSS of gravitational perturbations will be able to determine which satellites have cores. Accurate measurements of the density of each satellite, coupled with spectroscopically determined compositional information, will yield determinations of their bulk densities.

And finally, what was the origin of Phoebe? Is it a captured Kuiper Belt object or a more pedestrian asteroid? Is its surface material related to the dark material on either Iapetus or Hyperion?

With four years to investigate Saturn and its icy satellites, the Cassini–Huygens mission hopes to shed some light on the innumerable mysteries that remain about this solar system giant. Any answers that Cassini uncovers, of course, are equally likely to lead to even more questions.

CHAPTER 6

A Sphere of Influence

Saturn, its moons and its awesome rings sit inside an enormous cavity in the solar wind created by the planet's strong magnetic field. This "sphere of influence" of Saturn's magnetic field — called a magnetosphere — resembles a similar magnetic bubble surrounding Earth. The region is not at all spherical; rather, the supersonic solar wind, flowing at 300–1000 kilometers per second against Saturn's magnetic field, compresses the magnetosphere on the side facing the Sun and draws it out into a long magnetotail in the direction away from the Sun.

Inside the Magnetosphere

Inside Saturn's vast magnetospheric bubble is a mixture of particles, including electrons, various species of ions and neutral atoms and molecules, several populations of very energetic charged particles (like those in Earth's Van Allen Belts) and charged dust grains. The charged particles and dust grains all interact with both the steady and the fluctuating electric and magnetic fields present throughout the magnetosphere.

These ionized gases contain charged particles (electrons and ions) such as occur in the solar wind and planetary magnetospheres and are called plasmas. The steady fields can cause organized motions of the charged particles, creating large currents in the plasma.

Plasma behavior is more complex than that of neutral gases because, unlike neutral particles, the charged particles interact with each other electromagnetically as well as with any electric and magnetic fields present. The plasma's fluctuating fields (including wave fields) can "scatter" the charged particles in a manner similar to collisions in a neutral gas and cause a mixing of all the magnetospheric components.

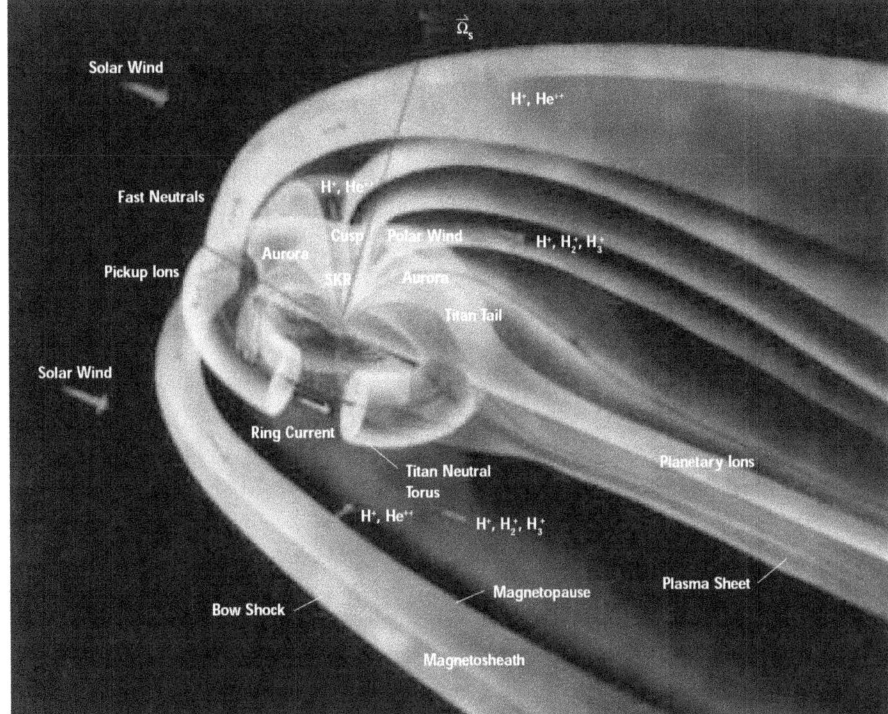

An artist's rendition of Saturn's immense magnetosphere. $\vec{\Omega}_S$ is the planet's rotation axis, closely aligned with the magnetic axis. [Image courtesy of Los Alamos National Laboratory]

Most of what we know about Saturn's magnetosphere comes from the brief visits by Pioneer 11 and Voyagers 1 and 2, but remote observations by the Hubble Space Telescope and other spacecraft have also provided us with intriguing information.

Magnetospheric Particle Sources

Saturn has a variety of sources for the particles in its magnetosphere. Particles can escape from any moon, ring or dust particle surface, or they can be "sputtered" off by energetic particles or even micrometeoroid impacts.

The primary particle sources are thought to be the moons Dione and Tethys. But, the solar wind, ionosphere, rings, Saturn's atmosphere, Titan's atmosphere and the other icy moons are sources as well. Recent Hubble Space Telescope results show large numbers of neutral hydrogen atoms (the neutral hydrogen cloud in the illustration above) throughout the magnetosphere that probably come from a number of these sources. It has even been proposed that water ions and molecules may form a dense "ionosphere" above Saturn's rings.

Recent Hubble Space Telescope results show large numbers of neutral hydrogen atoms throughout the magnetosphere that probably come from a number of the sources mentioned. Determining the relative importance of the varied sources in different parts of Saturn's space environment is a prime objective for the Magnetospheric and Plasma Science (MAPS) instruments aboard the Cassini spacecraft.

Neutral particles can escape from any moon, ring or dust particle surface — or they can be "sputtered" off by energetic particles or even micrometeoroid impacts. When these particles become ionized, they can excite electromagnetic waves with a frequency that can be used to determine their type. The icy rings absorb the energetic particles inward of the moon Mimas. Energetic particles can be created by processes within the magnetosphere or they can leak in from the solar wind. These and many other magnetospheric phenomena were seen by the three earlier spacecraft.

The mysterious "spokes" in the rings of Saturn, clearly seen in Voyager images, are probably caused by electrodynamic interactions between the tiny charged dust particles in the rings and the magnetosphere. Auroras, which exist on Saturn as well as Earth, are produced when trapped charged particles precipitating from the magnetosphere collide with atmospheric gases.

Despite many exciting discoveries, many more questions about the physical processes in Saturn's magnetosphere remain unanswered. This chapter examines the current state of knowledge about Saturn's magnetosphere and discusses the observations we expect to make with Cassini's instruments and the knowledge we expect to gain from forthcoming explorations.

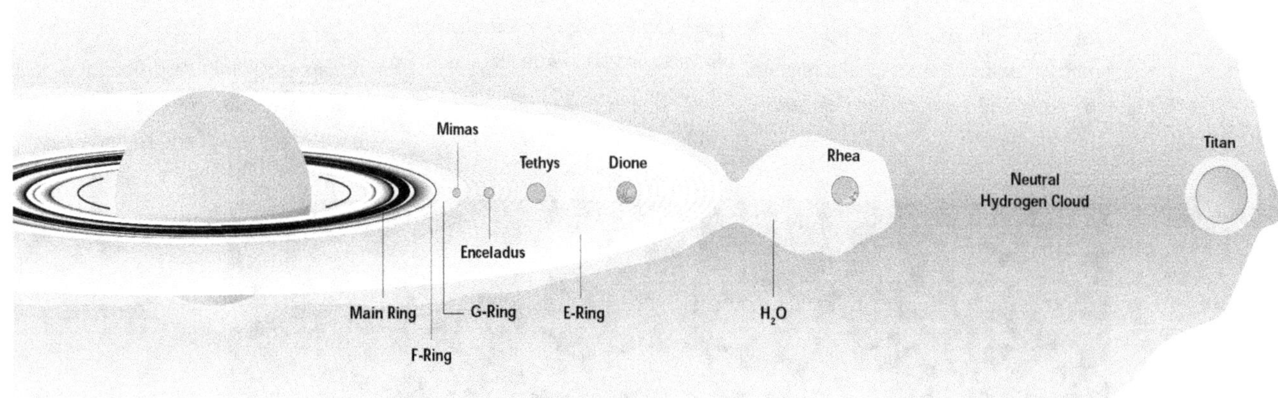

Sources of particles in Saturn's magnetosphere.

The MAPS Instruments

Coordinated observations are required from all the Magnetospheric and Plasma Science (MAPS) instruments aboard Cassini to fully understand Saturn's various dynamic magnetospheric processes. The Cassini Plasma Spectrometer will measure in situ Saturn's plasma populations including measurements of electron and ion species (H^+, H_2^+, He^{++}, N^+, OH^+, H_2O^+, N_2) and determine plasma flows and currents throughout the magnetosphere.

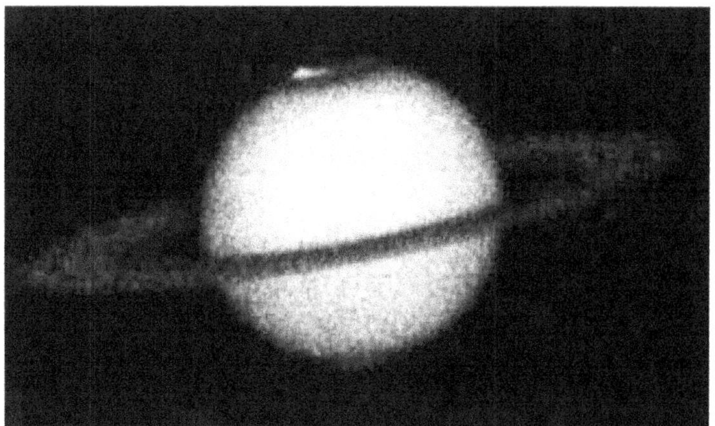

Saturn's aurora, imaged in the far ultraviolet by the Wide Field and Planetary Camera 2 aboard the Hubble Space Telescope. The aurora (the bright region near the pole) is caused by energetic charged particles exciting atoms in the upper atmosphere.

The Cosmic Dust Analyzer will make measurements of dust particles with masses of 10^{-19}–10^{-9} kilograms, determining mass, composition, electric charge, velocity and direction of incoming dust particles. Perhaps this instrument's most important capability will be measuring the chemical composition of incoming dust particles, making it possible to relate individual particles to specific satellite sources.

The Ion and Neutral Mass Spectrometer will measure neutral species and low-energy ions throughout the magnetosphere and especially at Titan. The Dual Technique Magnetometer will measure the strength and direction of the magnetic field throughout the magnetosphere. The first ever global images of Saturn's hot plasma regions will be obtained by the Magnetospheric Imaging Instrument, which will also measure in situ energetic ions and electrons.

The Radio and Plasma Wave Science instrument will detect the radio and plasma wave emissions from Saturn's magnetosphere, which will tell us about plasma sources and interactions in the magnetosphere. The Radio Science Instrument will measure the ionosphere of Saturn and search for ionospheres around Titan, the other moons and the rings. The Ultraviolet Imaging Spectrograph will map the populations of atomic hydrogen and weak emissions from neutrals and ions including auroral emissions.

The Magnetic Enigma

Saturn's magnetic field presents an enigma. Planetary fields such as those of Earth and Saturn can be approximated by a dipole, a simple magnetic field structure with north and south poles, similar to that produced by a bar magnet. Magnetic field measurements from the three previous flybys revealed a dipole-like field at Saturn with no (less than one degree) measurable tilt between Saturn's rotation and magnetic dipole axes. This near-perfect alignment of the two axes is unique among the planets. The Earth and Jupiter have dipole tilts of 11.4 and 9.6 degrees, respectively. The polarity of Saturn's magnetic dipole, like Jupiter's, is opposite to that of Earth.

There is a general consensus that the internal magnetic fields of the giant planets arise from dynamo action somewhere inside the planets' gaseous atmospheres. Of course, we do not really know what is inside Saturn or where the field is generated, although we have a number of theories. The inside of Saturn is probably quite exotic because of the great pressures caused by its large size. There may be a rocky (Earth-like) center with a molten core, but wrapped around this core we would expect to find layers of other uncommon materials (like liquid helium). The Saturn we see with telescopes and cameras is really only the cloud tops.

Although the measured field is symmetrical about the rotation axis, a number of observed phenomena can only be explained by an asymmetry in the magnetic field. Two examples are the occurrence of major emissions of Saturn kilometric radiation

(SKR), the principal radio emission from Saturn, at the presumed period of the planetary rotation and a similar variation in the formation of the spokes in the B-ring. The SKR observations can be explained by a magnetic anomaly in the otherwise symmetric field of less than five percent of the field at Saturn's surface (0.2 gauss), small enough to be imperceptible at the closest approach distances of the previous flybys of the Voyagers and Pioneer.

With magnetic field measurements made close to the planet over a wide range of latitudes and longitudes, the Dual Technique Magnetometer on Cassini will measure the details of the magnetic field and tell us more about Saturn's interior. The magnetometer will measure the strength and direction of the magnetic field throughout the magnetosphere, close to the planet where the field is dipolar and further from the planet where the field is non-dipolar due to distortion by current systems. The magnetometer will measure the field with sufficient accuracy to determine if it is indeed symmetrical. If so, the basic tenets of dynamo theory may need to be reexamined.

Solar Wind Interaction

A planetary magnetosphere forms when the magnetized solar wind (the supersonic, ionized gas that flows radially outward from the Sun) impinges upon a planet with a sufficiently large magnetic field. Like Earth and the other giant planets, Saturn has a strong magnetic field and an extensive magnetosphere. Although the morphology and dynamics of planetary magnetospheres vary according to the strength and orientation of their internal fields, magnetospheres share many common features.

Because the solar wind flow is almost always supersonic, a "bow shock" forms Sunward of the magnetosphere. The bow shock heats, deflects and slows the solar wind. Pioneer 11 made the first in situ measurements of Saturn's bow shock in 1979 when discontinuous jumps in solar wind parameters (magnetic field strength, density, temperature) were observed. Because of the variation in characteristics of the solar wind with distance from the Sun, by the time the orbit of Saturn is reached, the average Mach number, which determines the strength of the bow shock, is quite large. The bow shock of Saturn is a high Mach number shock similar to that of Jupiter and differs from the low Mach number shocks of the terrestrial planets. Saturn's bow shock provides a unique opportunity to study the structure of strong astrophysical shocks.

The magnetopause marks the boundary of the magnetosphere, separating the solar wind plasma and magnetospheric plasma. Between the bow shock and the magnetopause is a layer of deflected and heated solar wind material forming the magnetosheath. The boundaries move in and out in response to changing solar wind conditions. The average distance to the nose of the magnetopause at Saturn is roughly 20 R_S (R_S = one Saturn radius or 60,330 kilometers). These boundaries, shown in the image on the first page of this chapter, are of interest in understanding how energy from the solar wind is transferred to the planet to fuel magnetospheric processes.

Extensive observations of Earth's magnetosphere have demonstrated that solar wind energy is coupled into the magnetosphere primarily through a process called magnetic reconnection, in which field lines break and reconnect to change the magnetic topology. Similar processes must indeed occur at Saturn. Given the proper relative orientation of interplanetary and planetary magnetic fields on the sunward side of the magnetosphere, the field lines reconnect and a purely planetary magnetic field line (with both ends attached to the planet) becomes a field line with one end attached to the planet and the other end open to interplanetary space.

It is on these open field lines that form at high Saturn latitudes that energetic particles of solar, interplanetary or cosmic origin can enter the magnetosphere. These regions of the magnetosphere over the northern and southern poles are referred to as the polar caps. The open field lines are then pulled back by the drag of the diverted solar wind flow to make the magnetotail. Because charged particles and magnetic field lines are "frozen" together, this drives a tailward flow within the magnetosphere.

In the magnetotail, reconnection again occurs. Here, the magnetic field reverses direction across the tail's plasma sheet (a thin sheet of plasma located approximately in the planet's equatorial plane, where currents flow and particles are accelerated. The process of reconnection and opening of field lines on the sunward side of the magnetosphere is thus balanced by reconnection that closes field lines in the magnetotail. The newly closed field lines contract back toward the planet, pulling the plasma along and driving a circulation pattern, as shown in the figure below.

The process of reconnection on the Sunward side of the magnetosphere is thus closely coupled to processes that occur in the magnetotail. These processes are known to be strongly affected by the changing conditions in the solar wind. At Earth, reconnection processes can give rise to large, erratic changes in the global configuration of the magnetosphere referred to as geomagnetic storms. Cassini's MAPS instruments will investigate to see if similar magnetospheric storms occur at Saturn.

Voyager 1 made the first direct measurement of Saturn's magnetotail, finding it to resemble its terrestrial and Jupiter counterparts. The magnetotail was detected to be roughly 40 R_S in diameter at a distance 25 R_S downstream; it may extend hundreds of Saturn radii in the downstream solar wind. Understanding the processes that occur in the magnetotail is fundamental to understanding overall magnetospheric dynamics; coordinated measurements by the MAPS instruments during the deep tail orbits planned for the Cassini tour will contribute to that understanding. In turn, by understanding overall magnetospheric dynamics, scientists will gain insight into how Saturn's magnetosphere harnesses energy from the solar wind.

Current Magnetospheric Systems

Various large-scale current systems exist in Saturn's magnetosphere due to the collective motions of charged particles. Cross-tail currents flow from dusk to dawn in the plasma sheet located near the center of the magnetotail. An equatorial ring current distorts the magnetic field from its di-

SOLAR WIND CIRCULATION

Large-scale circulation driven by the solar wind as it occurs at Earth. An analogous process occurs at Saturn. The orientations of magnetic field lines and plasma flows are shown. When the interplanetary magnetic field is oriented southward, as shown, field lines reconnect at the nose of the magnetopause and then again in the magnetotail, driving the flows described in the chapter text.

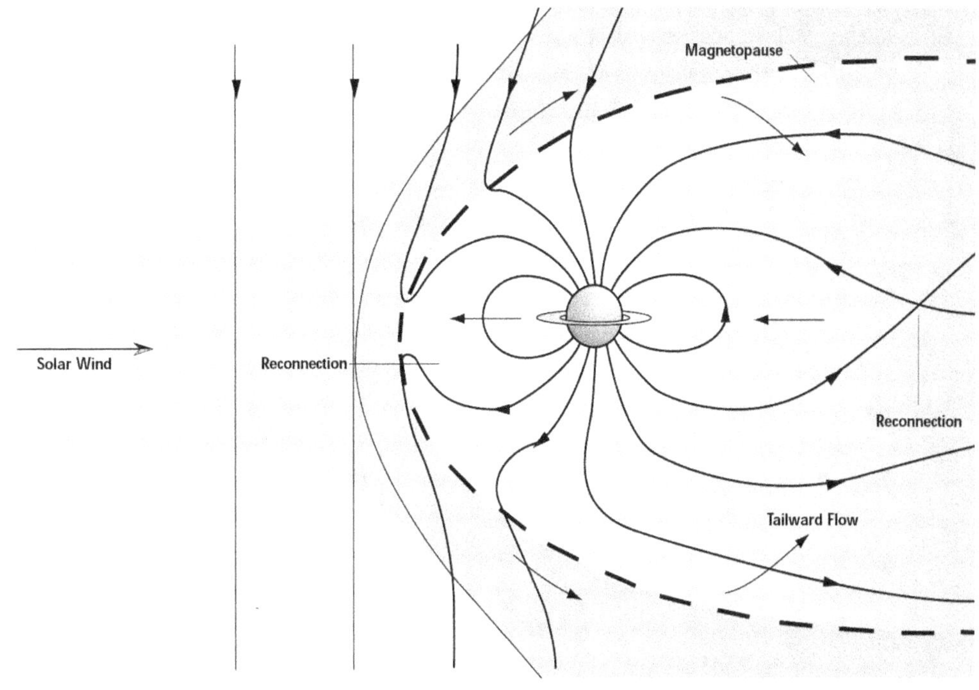

polar configuration, particularly in the outer magnetosphere where it stretches the magnetic field lines in the equatorial plane. This ring current, caused by electrons and ions drifting around the planet in opposite directions, is probably primarily due to the energetic particles discussed later in this chapter. The effect of this ring current is moderate when compared with Jupiter, however. Another major contribution to Saturn's total magnetic field comes from currents flowing in the magnetopause, which result from interaction with the solar wind.

Cassini's Dual Technique Magnetometer, measuring the magnetic field, and the Cassini Plasma Spectrometer, measuring the currents, will help map the current systems. These measurements, together with those taken by the other Cassini plasma instruments, will allow scientists to make a global model of Saturn's magnetic field throughout the magnetosphere.

Major Magnetospheric Flows

There are two primary sources of energy driving magnetospheric processes: the planet's rotation and the solar wind. Correspondingly, there are two types of large-scale plasma flow within the magnetosphere — corotation and convection. The nature of the large-scale circulation of particles in the magnetosphere depends on which source is dominant. At Earth, the energy is derived primarily from the solar wind; at Jupiter it is derived from the planet's rapid rotation rate. Saturn's magnetosphere is especially interesting because it is somewhere in between: both energy sources should play an important role.

Saturn's ionosphere is a thin layer of partially ionized gas at the top of the sunlit atmosphere. Collisions between particles in the atmosphere and the ionosphere create a frictional drag that causes the ionosphere to rotate together with Saturn and its atmosphere. The ionosphere, which extends from 1500 kilometers above the surface (defined as the visible cloud layer) to about 5000 kilometers, has a maximum density of about 10,000 electrons per cubic centimeter at about 2000–3000 kilometers.

The rotation of Saturn's magnetic field with the planet creates a large electric field that extends into the magnetosphere. The electromagnetic forces due to the combination of this electric field and Saturn's magnetic field cause the charged magnetospheric plasma particles to "corotate" (rotate together with Saturn and its internal magnetic field) as far out as Rhea's orbit (about nine R_S).

Convection, the other large-scale flow, is caused by solar wind pulling the magnetic field lines toward the tail. This leads to a plasma flow from day side to night side on open field lines and to a return flow from night side to day side on closed field lines (particularly near the equatorial plane).

On the dawn side, the corotation and convective flows will be in the same direction, but on the dusk side, they are opposing flows. The interaction of these flows may be responsible for some of the large variability observed in the outer magnetosphere. While at present we can only speculate about the consequences of these plasma flow patterns, we may expect some answers from investigations by Cassini's plasma instruments (especially the Cassini Plasma Spectrometer and the Magnetospheric Imaging Instrument).

Magnetospheric Plasma Regions

Saturn's magnetosphere can be broadly divided in two parts: a fairly quiet inner magnetosphere extending to about 12 R_S (beyond all moons except Titan), and an extremely variable hot outer magnetosphere. In both regions, the plasma particles are concentrated in a disk near the equatorial plane, where most plasma particle sources are located.

In their brief passages through Saturn's inner and outer magnetospheres, Pioneer and both Voyagers passed through several different plasma regions. The spacecraft observed a systematic increase in electron temperature with distance from Saturn, ranging from one electron volt (equivalent to a temperature of 11,600 kelvins) at four R_S in the inner magnetosphere and increasing to over 500 electron volts in the outer magnetosphere.

The thickness of the plasma disk increases with distance from Saturn. Inside about four R_S a dense (about 100 per cubic centimeter) popula-

SATURN'S MAGNETIC FIELD

Saturn's magnetic field. Field lines are shown for a dipole field model (solid line) and a model containing a dipole plus a ring current (dashed line). The stretching out of the field lines due to the ring current (shaded region) is moderate.

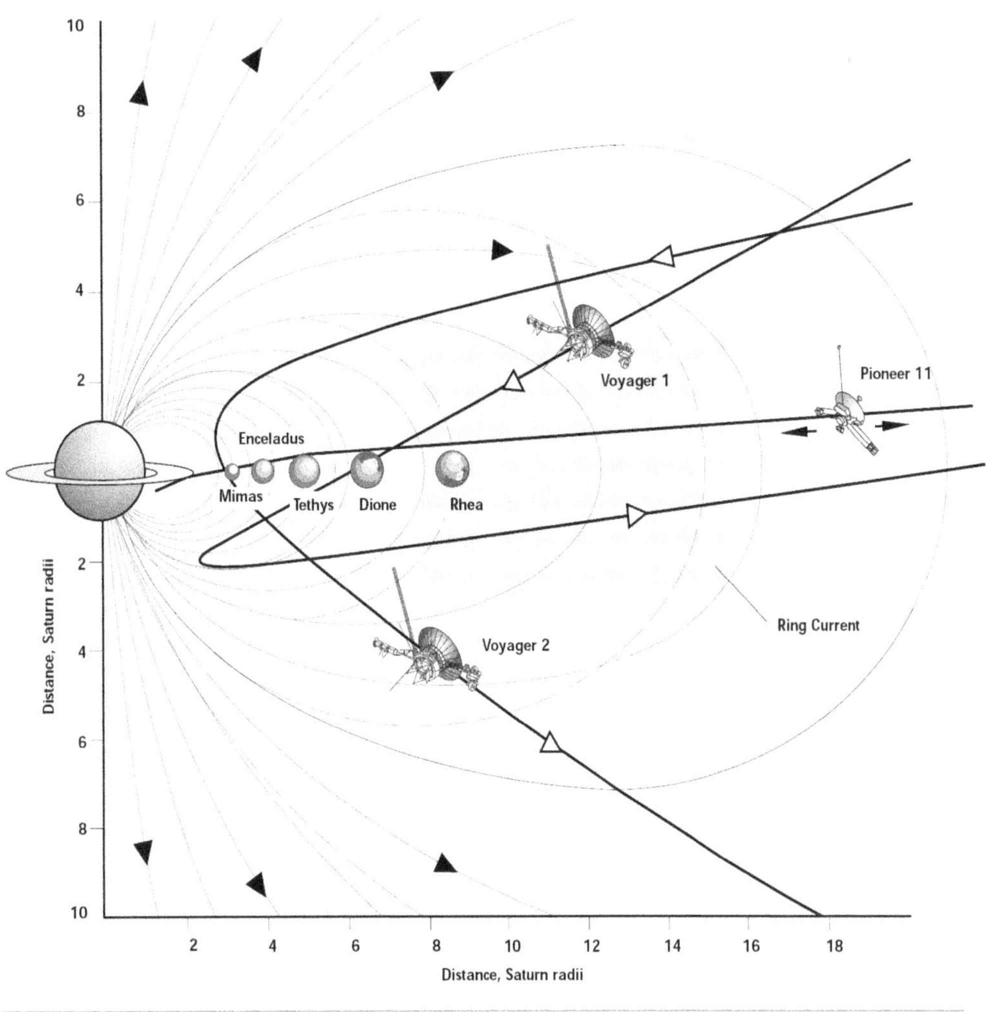

tion of low-energy ions and electrons is concentrated in a thin (less than 0.5 R_S) equatorial sheet. The low temperature is probably due to interactions with ring material; it has even been proposed that water ions and molecules may form a dense "ionosphere" above Saturn's rings.

In the inner magnetosphere, there is an oxygen-rich Dione–Tethys torus extending from four to about eight R_S, beyond Rhea's orbit. The icy surfaces of Dione and Tethys and other moons and rings in the magnetosphere are continually bombarded by both particles and solar radiation. Water molecules released by the bombardment form a disk-shaped cloud of water molecules and fragments of these molecules. The charged particle density in this region is a few particles per cubic centimeter and is composed of about 20 percent light ions (primarily hydrogen ions) and about 80 percent heavy ions with masses between 14 and 18 (species such as O^+ and OH^+).

In between Saturn's inner torus and outer magnetosphere is an extended equatorial plasma sheet of charged particles with densities between 0.1 and 2 particles per cubic centimeter. The inner edge of the sheet has "hot" (temperatures in the thousands of electron volts) ions and coincides with a vast cloud of neutral hydrogen, extending to 25 R_S, which probably escaped from the moon Titan, and other sources as well. Possibly, the hot ions are newly born ions from the neutral cloud that were heated by Saturn's rotational energy.

The Voyager spacecraft saw considerable variability in both the charged particle density and temperature in the outer magnetosphere on very short time scales. This has been interpreted as "blobs" of hot plasma interspersed with outward moving cold plasma and may be pieces of the plasma sheet that have broken off. The variability may also be due to dense "plumes" of hydrogen or nitrogen escaping from Titan that wrap around Saturn. Alternately, the variations may be caused by fluctuations in the solar wind, since both the outer magnetosphere and the magnetotail are thought to be the primary regions where solar wind energy enters the magnetosphere.

The dynamics, composition and sources of the outer magnetospheric plasma particles are not well understood. Investigation of this region is one important Cassini objective, so the MAPS instruments will make coordinated observations in this region. Until Cassini determines the composition here, the extent of the role of Titan in Saturn's outer magnetosphere will remain unknown.

In the inner region of the magnetosphere, most of the particles "corotate" with the planet. The corotation speed of charged particles differs from the speed of normal orbital motion (determined by gravity). Beyond approximately eight R_S, the charged particles lag behind the corotation speed by 10–30 percent. Here, the gravitational orbit velocity is much slower than the corotation speed; the lag is probably due to new ions born from the neutral hydrogen cloud or Titan's atmosphere that have not yet been brought up to Saturn's rotation rate (the corotation speed).

In the outer magnetosphere, the plasma rotation rate is about 30 percent lower than the corotation speed. When neutral particles from the moons or rings are ionized, they begin to move relative to the other

MAJOR MAGNETOSPHERIC FLOWS

Major flows in Saturn's magnetosphere. The solar wind flows in from the left; the magnetotail is to the right. Convection, a tailward plasma flow, is caused by the solar wind dragging magnetic field lines past the planet. Corotation, magnetospheric rotation at the rate of Saturn, is caused by the corotation of Saturn's ionosphere.

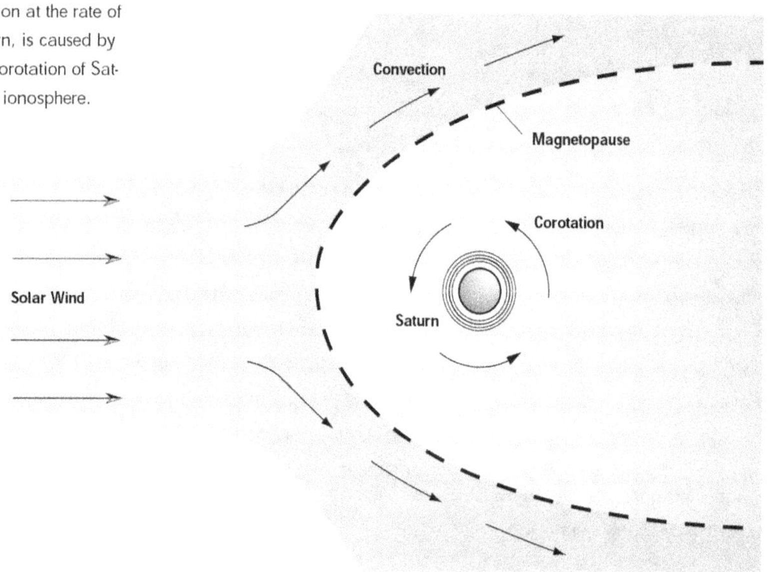

REGIONS OF SATURN'S MAGNETOSPHERE

Regions of Saturn's magnetosphere. The various plasma regions — inner torus, extended plasma sheet, variable outer magnetosphere, etc. — are shown in relationship to the location of the moons and the magnetosheath. The temperature is indicated by the color scale, going from cold (blue) to hot (pink). Note the asymmetry: The left side shows the noon magnetosphere (that portion closest to the Sun). [Based on Sittler, et al., 1983]

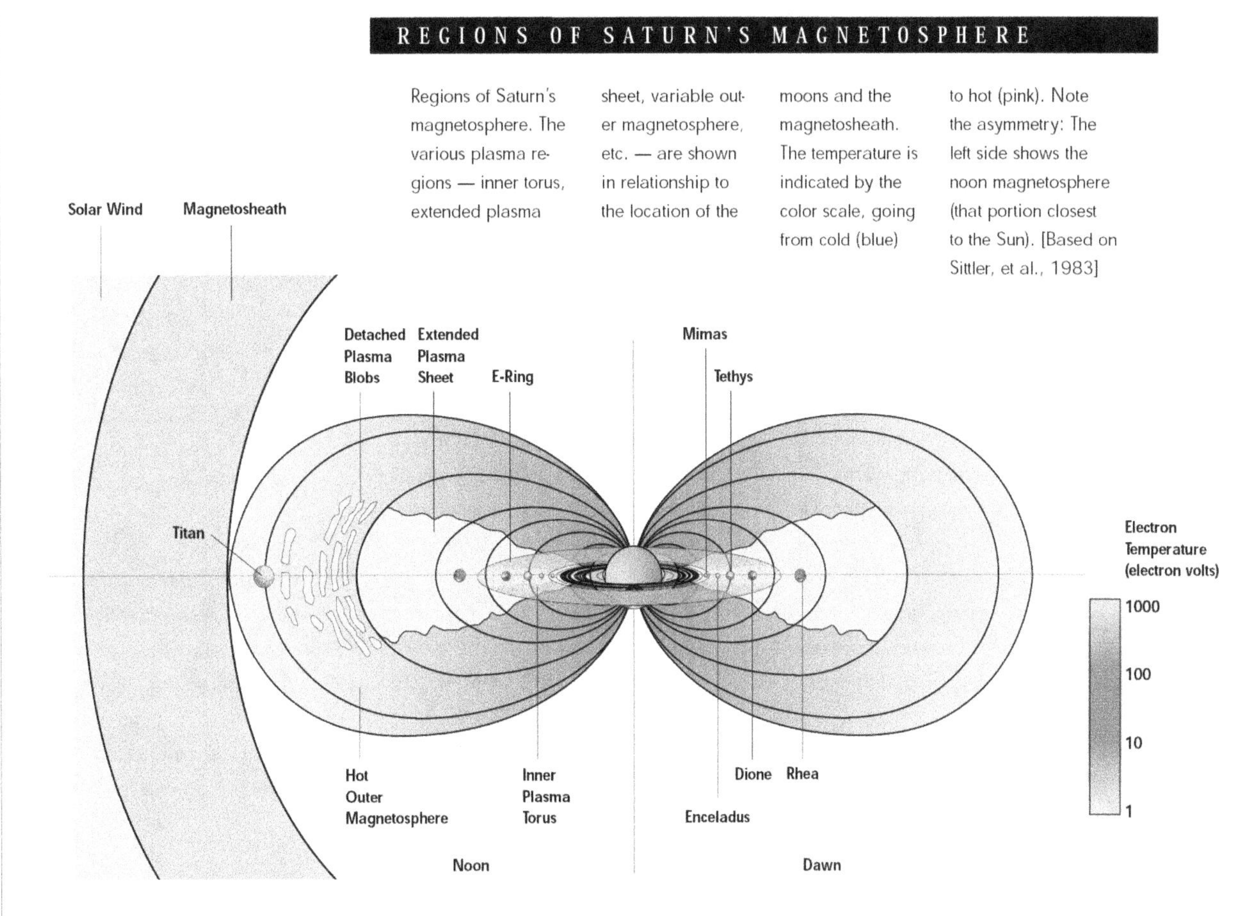

neutral particles because of the difference in the orbital speed and the corotation speed. Since these new ions add to the mass of the corotating plasma population, they can slow it down, as suggested by the observations. Outward motion of plasma from the inner magnetosphere may also contribute to slowing it down. Cassini's MAPS instruments will investigate the relative importance of these two effects on the corotation rate in the outer magnetosphere.

Energetic Particle Populations

Saturn's magnetosphere, like that of other planets, contains populations of highly energetic particles similar to those in Earth's Van Allen radiation belts (kilo electron volt to mega electron volt energies). These particles are trapped by Saturn's strong magnetic field.

In a uniform magnetic field, charged particles move in helical orbits along magnetic field lines. In Saturn's dipolar magnetic field, the field strength along a field line increases toward the planet. At some point determined by the particle speed and the magnetic field strength, the particle is "reflected" or "mirrored" and it reverses direction along the same field line. A "trapped" charged particle moves in such an orbit in Saturn's field, bouncing back and forth along a single magnetic field line. The radiation belts are made up of energetic particles moving in such orbits. Collisions with neutral particles or interactions with the fluctuating electric and magnetic fields in the plasma can change a charged particle's orbit.

Voyager 2 data showed Saturn's magnetosphere to be populated largely by low-energy (tens of electron volts) electrons in the outer regions with more energetic electrons dominating further inward. Substantial fluxes of high-energy protons were observed inside the orbits of Enceladus and Mimas, forming the hard core of the radiation belts. Pioneer 11 investigators concluded that these protons probably originated from the interaction of cosmic rays with Saturn's rings.

The origin of these and other energetic particles is unclear and will be investigated by Cassini's MAPS instruments. In particular, the Magnetospheric Imaging Instrument will make in situ measurements of energetic ions and electrons. Some energetic ions such as helium and carbon may originate in the solar wind, but others may come from lower energy particles that are energized in Saturn's magnetosphere.

The energetic particles drifting in the dipole-like magnetic field create the ring currents discussed above. Charged particles moving in trapped particle orbits along dipole field lines also drift in circles around the planet. Electrons and ions drift in opposite directions and this causes the ring current discussed previously in this chapter.

Measurements of energetic particles indicate that the satellites of Saturn play an important role in shaping their spatial distributions. In the inner region of the magnetosphere, charged particles undergo significant losses as they diffuse inward and are swept up by collisions with the satellites.

Some of the energetic ions undergo collisions with the surrounding neutral gases that result in the exchange of an electron, producing a population of fast or energetic neutral atoms in Saturn's magnetosphere. Cassini's Magnetospheric Imaging Instrument will use these energetic neutral atoms as if they were photons of light to make global images and study the overall configuration and dynamics of Saturn's magnetosphere. The instrument will obtain the first global images of Saturn's hot plasma regions with observations of features such as Saturn's ring current and Titan's hydrogen torus. Cassini will be the first spacecraft to carry an instrument to image the magnetosphere using energetic neutral atoms.

Polar Region Interactions

Aurora. Most energetic particles bounce back and forth along field lines in trapped particle orbits. If, however, the mirror point is below the top of the atmosphere, the particle can deposit its energy in the up-

CHARGED PARTICLE ORBITS

Charged particle orbits in a magnetic field. Left: in a uniform field, charged particles are tied to field lines and move along them in helical orbits. Right: in a dipole-like field, trapped charged particles move in helical orbits along field lines, but at some point "mirror" or "reflect," leading to a bounce motion along the field line. Charged particles in such trapped orbits also drift in circles around the planet due to the inhomogenous magnetic field. Ions drift in one direction and electrons in the other, leading to a ring current that modifies the planetary magnetic field.

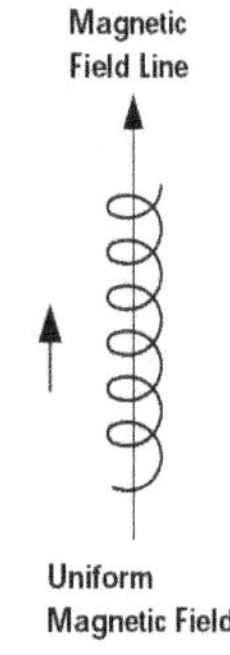

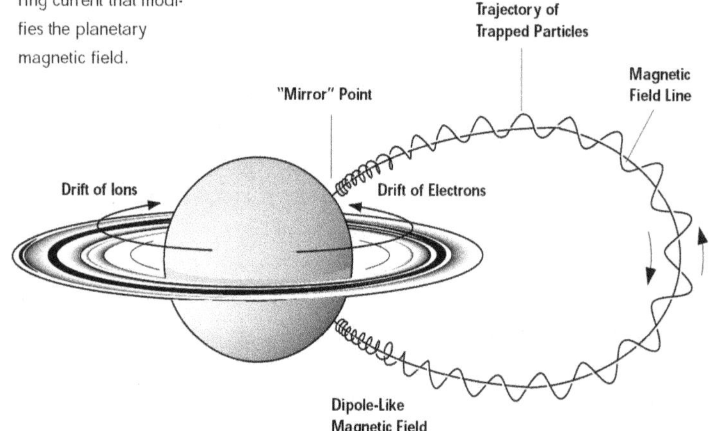

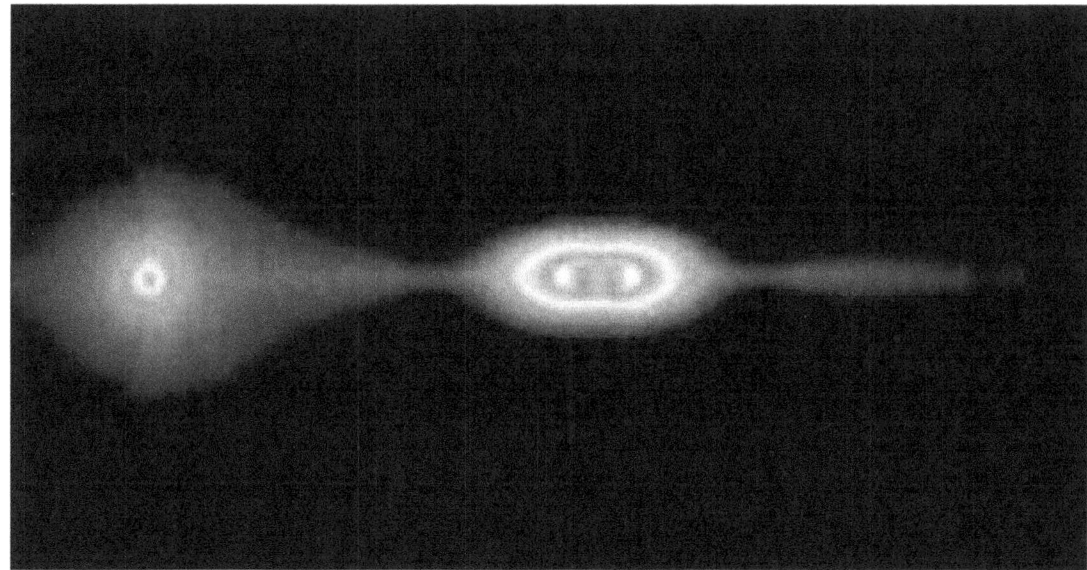

Energetic neutral imaging. Simulation of an energetic neutral atom (ENA) image of the type that will be obtained by Cassini's Magnetospheric Imaging Instrument. The Saturn magnetosphere appears close to the center of the image and Titan is on the left.

per atmosphere. Energetic particles reaching the atmosphere create the auroral emission by exciting gases in the upper atmosphere (molecular and atomic hydrogen lines in the case of Saturn; oxygen and nitrogen in Earth's atmosphere).

Saturn's aurora was first detected by the Voyager ultraviolet spectrometer. While is it not clear which magnetospheric particles (electrons, protons or heavy ions) create the aurora, it is clear that planets with higher fluxes of energetic particles have stronger auroral emissions. Cassini's MAPS instruments will make coordinated studies of Saturn's aurora, with the Ultraviolet Imaging Spectrometer providing images.

Saturn Kilometric Radiation. For about 20 years prior to the Voyager visits to Saturn, radio astronomers had been searching for Saturn's radio emissions. We now know that Saturn is a much weaker radio source than Jupiter. Confirmation of radio emissions from Saturn came only when Voyager 1 approached within three astronomical units of the planet.

Saturn emits most strongly at kilometric wavelengths. Like the radio emission of other planets, Saturn kilometric radiation (SKR) comes from the auroral regions of both hemispheres and the radio beams are fixed in Saturn's local time. However, the emitting regions are on the night side for Earth and on the dayside for Saturn. The emission appears to come from localized sources near the poles — one in the north and one in the south — that "light up" only when they reach a certain range of local times near Saturn's noon.

The periodicity of these emissions is about 10 hours, 39 minutes, assumed to be the rotation rate of Saturn's conducting core. This is somewhat longer than the atmospheric rotation rate of 10 hours, 10 minutes observed at the cloud tops near the equator. The periodicity in SKR emission is unexpected for a planet with such a symmetrical magnetic field. Possibly, a magnetic anomaly exists that allows energetic electrons to penetrate further down into the polar region (at some point) and here the SKR radiation is generated at the electron's natural frequency of oscillation.

Based on the local time of emission, the source energy for the SKR appears to be the supersonic solar wind and, in fact, changes in the solar wind strongly control the SKR power. For example, a solar wind pressure increase by a factor of about 100 results in an increase by a factor of

THE FOURTH STATE OF MATTER

Plasma is the fourth state of matter. A plasma is an ionized gas containing negatively charged electrons and positively charged ions of a single or many species; it may also contain neutral particles of various species. Examples are the Sun, the supersonic solar wind, Earth's ionosphere and the interstellar material. Plasmas behave differently from neutral gases; the charged particles interact with each other electromagnetically and with any electric and magnetic fields present. The charged particles also create and modify the electric and magnetic fields. In a highly conducting plasma, the magnetic field lines move with (are "frozen to") the plasmas. The X-ray image here shows a million-degree plasma, the solar corona, which is the source of the supersonic solar wind plasma that pervades the solar system. [Image from the Yohkoh satellite]

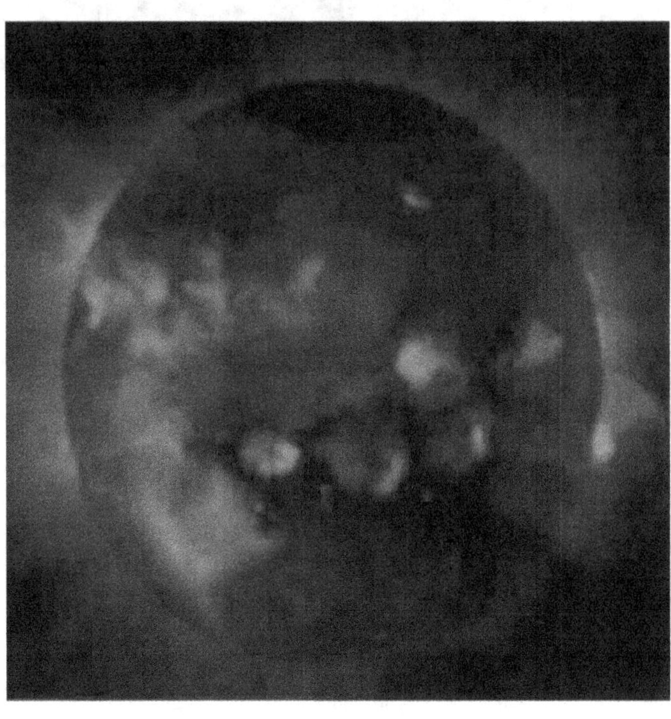

about 10 in the SKR power. For a period of about two to three days following the Voyager 2 encounter, no SKR emission was detected. It is thought that since Saturn was immersed in Jupiter's long magnetotail at this time, the planet's magnetosphere was shielded from the solar wind. One of the main objectives of the Radio and Plasma Wave Science instrument aboard Cassini is to make measurements of the SKR, study its variation with variations in the solar wind and map the source region.

Other Magnetospheric Emissions

We are all familiar with waves in a vacuum (electromagnetic waves) and waves in a gas and fluid (electromagnetic waves, sound waves, gravity waves). A magnetized plasma supports all these waves and more. Because of the electromagnetic interactions between charged plasma particles and the magnetic field, new types of waves can propagate that have no counterpart in a neutral gas or fluid. Waves in the magnetosphere can be produced via various processes, for example by ionization of atmospheric neutral atoms in the magnetospheric plasma or by currents flowing between different plasma populations.

These waves, as well as other types of waves (Alfvén waves, magnetosonic waves and ion and electron cyclotron waves, to name a few) can propagate in a plasma and be detected by sensors such as the magnetic and electrical antennas of Cassini's Radio and Plasma Wave Science instrument. These waves are trapped within the magnetosphere and thus can only be sampled inside it. Saturn produces a variety of radio and plasma wave emissions from narrow (single frequency) and bursty to broadband (several frequencies) and continuous. The primary goal of the RPWS instrument is to study these wave emissions. As mentioned, neutral atoms from various sources supply Saturn with magnetospheric plasma. As they do, they leave a "signature" in the plasma waves that can be used to determine their species.

Emissions in magnetospheres are waves of the plasma driven to large amplitudes by magnetospheric processes that tap some reserve of free energy. There are many modes, interactions and energy reserves; the emissions are studied to help discover the interactions and energy sources driving them.

Free Energy Sources. We have seen how energetic particles from the solar wind are one source of energy. Both nonuniform and nonthermal plasma distributions represent additional sources of free energy. Generally, waves that grow at the expense of a nonthermal or nonuniform feature interact back on the plasma distribution to try to eliminate the nonuniform or nonthermal feature. For example, the Pioneer 11 magnetometer saw low-frequency waves associated with Dione; these have been interpreted as ion cyclotron waves, apparently resonant with oxygen ions. These waves were probably generated by newly born oxygen ions, created from Dione's ice as a sputtering product, interacting with the corotating magnetosphere plasma and tapping the energy in the plasma rotation.

The waves generated by these new ions then act to thermalize their highly nonthermal distribution. A modulation of the radio emission was also

THOSE SURPRISING SPOKES

The "spokes" in Saturn's rings were first seen from Earth, but Voyager observations allowed the first study of how these surprising features evolve. The spokes are cloud-like distributions of micrometer-sized particles that occasionally appear in the region from approximately 1.75 R_S to 1.9 R_S. Voyager saw spokes form radially over thousands of kilometers in less than five minutes. Subsequent Keplerian motion (motion due to gravity) changes these spokes into "wedges." The images shown here form a time sequence from upper left to lower right. Most likely, these nonradial features result from interactions of tiny charged ring dust particles with the electromagnetic fields and/or charged particles in the magnetosphere. Moreover, the spokes occur preferentially at the same longitude and the same periodicity as the Saturn kilometric radiation, albeit at different local times, suggesting a relationship to the magnetic anomaly. Cassini's MAPS and imaging instruments will make coordinated studies of the formation and evolution of the spokes.

CASSINI'S MAPS INSTRUMENTS

Instrument	Objective
Cassini Plasma Spectrometer	Measures composition, density, velocity and temperature of ions and electrons
Cosmic Dust Analyzer	Measures flux, velocity, charge, mass and composition of dust and ice particles from 10^{-16}–10^{-6} grams
Dual Technique Magnetometer	Measures the direction and strength of the magnetic field
Ion and Neutral Mass Spectrometer	Measures neutral species and low-energy ions
Magnetospheric Imaging Instrument	Images Saturn's magnetosphere using energetic neutral atoms, and measures the composition, charge state and energy distribution of energetic ions and electrons
Radio and Plasma Wave Science	Measures wave emissions as well as electron density and temperature
Radio Science Instrument	Measures the density of Saturn's ionosphere
Ultraviolet Imaging Spectrograph	Measures ultraviolet emissions to determine sources of plasma in Saturn's magnetosphere

associated with the orbital phase of Dione, raising the possibility that Dione is venting gases. Plasma waves can also scatter particles into orbits, taking them down into the upper atmosphere, where they drive auroral processes. Although the RPWS instrument is the primary detector of plasma waves, the causes and effects of the plasma waves are seen in measurements by Cassini's other MAPS instruments and coordinated observations of wave phenomena will be important in understanding the sources and sinks of magnetospheric plasma and dynamic processes in general.

Atmospheric Lightning. Lightning in Saturn's atmosphere is thought to cause the unusual emissions designated Saturn electrostatic discharges (SED). These are short, broadband bursts of emission apparently coming from very localized regions (presumably atmospheric storms). It was determined that the source acts like a searchlight and is not fixed relative to the Sun, as is the case for SKR emissions.

A 10 hours, 10 minutes periodicity was seen in the emissions by Voyager 1, quite different from the 10 hours, 39 minutes periodicity of the SKR emissions. From Voyager imaging results, the rotational period of the equatorial cloud tops had also been measured at 10 hours, 10 minutes, consistent with the interpretation of the source as lightning. Cassini will further investigate the nature of these bursts, which give potential insight into Saturn's atmospheric processes, the planet's "weather." The Cassini RPWS instrument will make measurements of SKR emissions, electromagnetic emissions from lightning and SED as well. Measurements by the Cassini MAPS instruments will enhance an understanding of Saturn's complex and fascinating magnetosphere.

CHAPTER 7

Planning a Celestial Tour

Space mission design aims to maximize science return within the constraints and limitations imposed by the laws of nature and the safe, reliable operation of the spacecraft. Major mission design variables that may influence science return from the Cassini–Huygens mission are design and selection of the interplanetary trajectory, design and selection of spacecraft orbits and the use of the spacecraft and instruments in making observations and returning data to Earth. This chapter discusses the role of planetary swingbys in achieving the Cassini–Huygens trajectory to Saturn, along with the spacecraft's activities during this time.

The Interplanetary Mission

The Cassini–Huygens mission uses trajectories requiring planetary swingbys to achieve the necessary energy and orbit shaping to reach Saturn. The primary trajectory for Cassini is a Venus–Venus–Earth–Jupiter Gravity Assist (VVEJGA) transfer to Saturn.

As the name implies, the VVEJGA trajectory makes use of four gravity-assist planetary swingbys between launch from Earth and arrival at Saturn. The use of planetary gravity assists reduces launch energy requirements compared to other Earth–Saturn transfer modes, and allows the spacecraft to be launched by the Titan IVB/Centaur. Direct Earth–Saturn transfers with this launch vehicle are not possible for Cassini–Huygens.

The nominal launch period of the primary mission opens on October 6, 1997, and closes on November 4, 1997, providing a 30-day launch period. A contingency launch period is extended beyond the nominal launch period to November 15, 1997, to increase the chances of mission success — although possibly degrading to some extent the scientific accomplishments of the nominal mission. The

THE EASY WAY TO FLY

The Venus–Venus–Earth–Jupiter Gravity Assist (VVEJGA) trajectory that Cassini–Huygens will take requires a deep space maneuver between the two Venus swingbys. This maneuver reduces the spacecraft orbit perihelion (the closest point with respect to the Sun) and places it on the proper course to encounter Venus for a second time in June 1999. After the Earth swingby in August 1999, the Cassini–Huygens spacecraft will be on its way to the outer planets, flying by Jupiter in late December 2000. The fortuitous geometry of VVEJGA provides the unique opportunity of a double gravity-assist swingby, Venus 2 to Earth, within 56 days, reducing the total flight time to Saturn to under seven years. The scientific information obtained during the interplanetary cruise phase is limited primarily to gravitational wave searches during three successive Sun oppositions, beginning in December 2001.

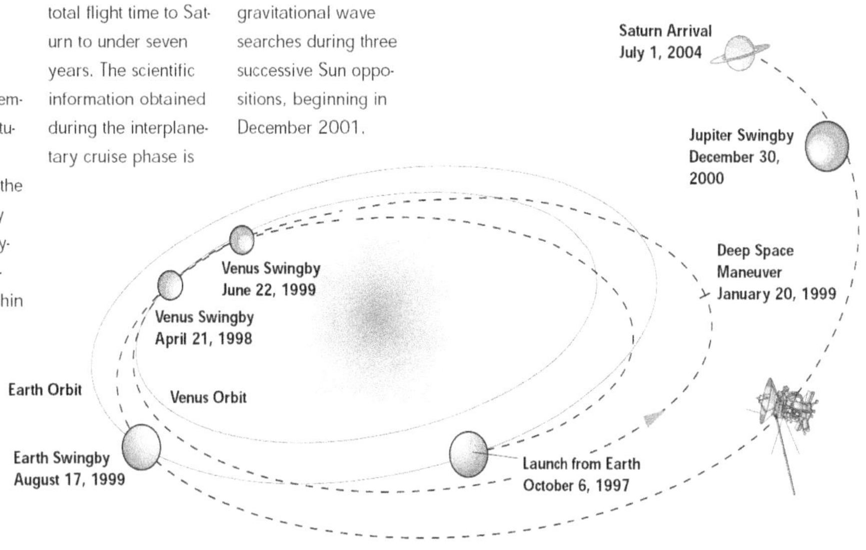

opening and closing of the nominal launch period are chosen such that the launch vehicle's capabilities are not exceeded and the mission performance and operational requirements are met.

After launch, the trajectory is controlled through a series of trajectory correction maneuvers designed to correct errors in the planetary swingbys. Each planetary swingby has the effect of a large maneuver on the trajectory of the spacecraft. For the prime trajectory, the four swingbys can supply the equivalent of over 20 kilometers per second of Sun-relative speed gains — an amount not achievable using conventional spacecraft propulsion.

Typically, interplanetary swingbys are controlled using two approach maneuvers and one departure maneuver, with additional maneuvers added at critical points. About 20 maneuvers will be needed to deliver the spacecraft from launch to Saturn. The controllers' knowledge of the actual location of the spacecraft is obtained using measurements made from the Deep Space Network. This system of maneuvers is capable of delivering the spacecraft to various planetary swingbys with an accuracy varying from about five kilometers at Earth to 150 kilometers at Jupiter. The final approach to Saturn is predicted to have an accuracy of about 30 kilometers.

Huygens Probe Mission

Based on the primary launch opportunity, Cassini–Huygens will arrive at Saturn on July 1, 2004. On arrival, the Orbiter will make a close flyby and execute a Saturn orbit insertion

MAJOR MISSION MILESTONES
CASSINI-HUYGENS

Milestone	Date
Launch	10/06/1997
Venus 1 Swingby	04/21/1998
High-Gain Antenna Opportunity	12/16/1998 – 01/10/1999
Deep Space Maneuver	01/20/1999
Venus 2 Swingby	06/22/1999
Earth Swingby	08/17/1999
Asteroid Belt Crossing	12/12/1999 – 04/10/2000
High-Gain Antenna Earth-Pointing	02/01/2000
Jupiter Swingby	12/30/2000
Gravity Wave Experiment	11/26/2001 – 01/05/2002
Conjunction Experiment	06/06/2002 – 07/06/2002
Cruise Science On	07/02/2002
Gravity Wave Experiment	12/06/2002 – 01/15/2003
Phoebe Flyby	06/11/2004
Saturn Orbit Insertion	07/01/2004
Periapsis Raise	09/25/2004
Probe Separation	11/06/2004
Orbiter Deflection	11/08/2004
Probe Titan Entry	11/27/2004
Titan 1 Flyby	11/27/2004
Titan 2 Flyby	01/14/2005
Titan 3 Flyby	02/15/2005
Enceladus 1 Flyby	03/09/2005
Titan 4 Flyby	03/31/2005
Titan 5 Flyby	04/16/2005
Enceladus 2 Flyby	07/14/2005
Titan 6 Flyby	08/22/2005
Titan Flybys Continue	2005 – 2008
Mission Ends	07/01/2008

(SOI) propulsive maneuver to initiate a highly elliptical, 148-day orbit around the planet. This orbit will set up the geometry for the first encounter with Titan for the Huygens Probe mission, currently planned for November 27, 2004.

On November 6, 2004, approximately 22 days before the first Titan flyby, the Huygens Titan Probe will be released from the Orbiter. The Orbiter will turn to orient the Probe to its entry attitude, spin it up to about 7.29 revolutions per minute and release it with a separation velocity of about 0.33 meter per second. At least two navigational maneuvers will be performed before separation to ensure accurate targeting for atmospheric entry. Two days after separation, the Orbiter will perform an Orbiter deflection maneuver (ODM) to ensure that the Orbiter will not follow the Probe into Titan's atmosphere, and to establish the proper geometry for the Probe data relay link.

The Orbiter is targeted to similar aimpoint conditions at the second Titan flyby to permit a contingency Probe mission opportunity if anything prevents the Probe from being delivered on the first flyby.

Following completion of the predicted descent, the Orbiter will continue to listen to the Probe for 30 minutes, in the event the Probe transmissions continue after landing. The longest predicted descent time is 150 minutes.

When Probe data collection is completed, that data will be write-protected on each of the Orbiter's solid-state recorders. The spacecraft will then turn to view Titan with optical remote-sensing instruments, until about one hour after closest approach.

Soon after closest approach, the Orbiter will turn the high-gain antenna toward Earth and begin transmitting the recorded Probe data. The complete, four-fold redundant set of Probe data will be transmitted to Earth twice, and its receipt verified, before the write protection on that portion of the recorder is lifted by ground command — marking Probe mission completion.

Saturn Orbiter Tour

The tour phase of the mission will begin at Probe mission completion and ends four years after the SOI. The reference tour described here, called Tour T18-3, consists of 74 orbits of Saturn with various orientations, orbital periods ranging from seven to 155 days and Saturn-centered periapsis radii ranging from about 2.6 to 15.8 R_S (Saturn radii).

Orbital inclinations with respect to Saturn's equator range from zero to 75 degrees, providing opportunities for ring imaging, magnetospheric coverage and radio (Earth), solar and stellar occultations of Saturn, Titan and the ring system.

JUST DROPPING BY

Shown here is the Cassini–Huygens flight path for the approach to Saturn, the Saturn orbit insertion maneuver, the initial Saturn orbit and the approach to the first Titan encounter. A propulsive maneuver — called the periapsis raise maneuver — during the initial orbit, near apoapsis (the farthest point from Saturn), raises periapsis (the nearest point to Saturn in the orbit) to correctly target the Orbiter for the first Titan flyby. If any problem arises with the Orbiter, Probe or ground system that prevents execution of the Probe mission at the first Titan flyby, mission controllers can decide to have the spacecraft fly by Titan without releasing the Probe, delaying the Probe's mission till the second Titan flyby on the second orbit of Saturn. The second Titan flyby is currently planned for January 14, 2005.

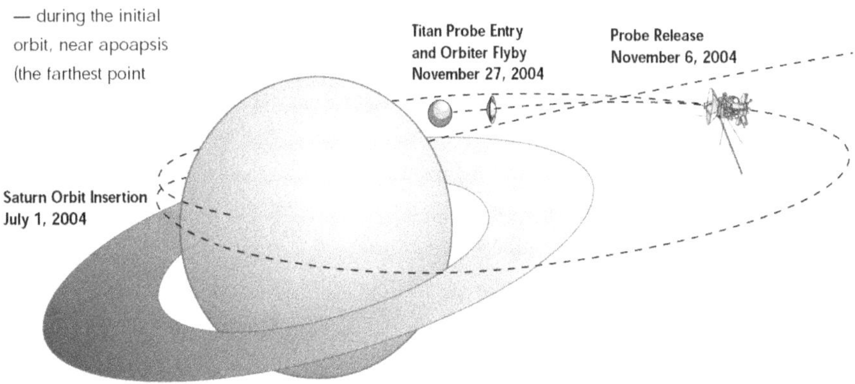

A total of 43 Titan flybys occur during the reference tour. Of these, 41 have flyby altitudes less than 2500 kilometers and two have flyby altitudes greater than 8000 kilometers. Titan flybys are used to control the spacecraft's orbit about Saturn as well as for Titan science acquisition. Our reference tour also contains seven close flybys of icy satellites and 27 additional distant flybys of icy satellites within 100,000 kilometers.

Close Titan flybys could make large changes in the Orbiter's trajectory. For example, a single close flyby of Titan can change the Orbiter's Saturn-relative velocity by hundreds of meters per second. For comparison, the total such change possible from the Orbiter's main engine and thrusters is about 500 meters per second for the entire tour. The limited amount of propellant available for tour operations is used only to provide small trajectory adjustments necessary to navigate the Orbiter, or to turn it in order to obtain science observations or to communicate with Earth.

Titan is the only satellite of Saturn massive enough to use for orbit control during a tour. The masses of the other satellites are so small that even close flybys (within several hundred kilometers) can change the Orbiter's trajectory only slightly. Consequently, Cassini tours consist mostly of Titan flybys. This places restrictions on how the tour must be designed. Each Titan flyby must place the Orbiter on a trajectory that leads back to Titan. The Orbiter cannot be targeted to a flyby of a satellite other than Titan unless the flyby lies almost along a return path to Titan. Otherwise, since the gravitational influence of the other satellites is so small, the Orbiter will not be able to return to Titan — and

STEERING BY THE STARS

The Cassini–Huygens tour is navigated with a combination of Doppler data and images against a star background. Each swingby of Titan modifies the spacecraft's trajectory — on average, each Titan encounter will provide the equivalent of a 770-meters-per-second spacecraft maneuver. (The icy satellites are too small to significantly modify the trajectory.) In a tour with 45 targeted encounters, the spacecraft will perform about 135 maneuvers. Shown here is a three-maneuver sequence to control the trajectory from one Titan encounter to the next. The "cleanup" maneuver corrects for errors in the previous flyby. The sequence uses between eight and 11 meters per second, per encounter. The plan is to allow some variation in the actual Titan swingby conditions that will reduce the amount of propellant required to accomplish the tour. The use of both radiometric and optical data results in a Titan delivery accuracy of 10 kilometers or less.

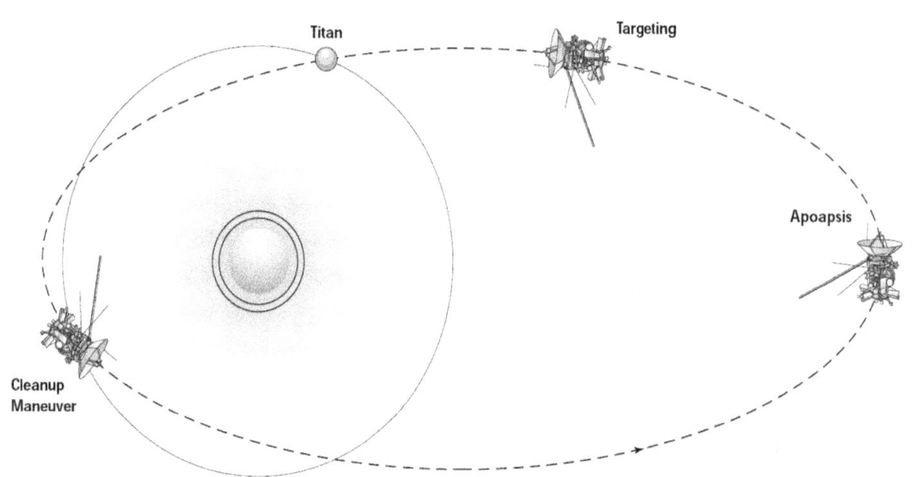

the tour cannot continue. Of course, the large number of Titan flybys will produce extensive coverage of Titan.

Tour Terminology

"Rev" numbers ranging from 1 to 74 are assigned to each tour orbit, assuming each orbit begins at apoapsis. The partial orbit from SOI to the first apoapsis is Rev 0. The rev number is incremented at each succeeding apoapsis. Two or more revs about Saturn may occur between successive Titan flybys; therefore, there is no correspondence between rev number and flyby number. Titan flybys are numbered consecutively, as are the targeted flybys of icy satellites. For example, the first flyby of Enceladus is Enceladus 1; the first of Rhea is Rhea 1.

"Orbit orientation" is the angle measured clockwise at Saturn from the Saturn–Sun line to the apoapsis. This is an important consideration for observations of Saturn's magnetosphere and atmosphere. The time available for observations of Saturn's lit side decreases as the orbit rotates toward the anti-Sun direction. Arrival conditions at Saturn fix the initial orientation at about 90 degrees. Due to the motion of Saturn around the Sun, the orbit orientation increases with time, at a rate of orientation of about one degree per month, which over the four-year tour results in a total rotation of about 48 degrees clockwise (as seen from above Saturn's north pole). Period-changing targeted flybys that rotate the line of apsides (orbital points nearest or farthest from the center of Saturn) may be used to add to or subtract from this drift in orbit orientation.

The "petal" plot on the facing page shows how targeted flybys combine with orbit drift to rotate the orbit from the initial orientation clockwise most of the way around Saturn to near the Sun line. In the coordinate system used in this diagram, the direction to the Sun is fixed. Encounters of satellites occur either inbound (before Saturn-centered periapsis) or outbound (after Saturn-centered periapsis).

A "targeted flyby" is one where the Orbiter's trajectory has been designed to pass through a specified aimpoint (latitude, longitude, altitude) at closest approach. At Titan, the aimpoint is selected to produce a desired change in the trajectory using the satellite's gravitational influence. At targeted flybys of icy satellites, the aimpoint is generally selected to optimize the opportunities for scientific observations, since the gravitational influence of those satellites is small. However, in some cases the satellite's gravitational influence is great enough to cause unacceptably large velocity-change penalties for a range of aimpoints, which makes it necessary to constrain the range of allowable aimpoints to avoid the penalty.

If the closest approach point during a flyby is far from the satellite, or if the satellite is small, the gravitational effect of the flyby can be small enough that the aimpoint at the flyby need not be tightly controlled. Such flybys are called "nontargeted." Flybys of Titan at distances greater than 25,000 kilometers — as well as flybys of satellites other than Titan at distances of greater than a few thousand kilometers — are considered nontargeted flybys.

Flybys of satellites other than Titan at distances up to a few thousand kilometers must be treated as targeted flybys to achieve science objectives, even though the satellite's gravitational influence is small. Opportunities to achieve nontargeted flybys of smaller satellites will occur frequently during the tour and are important for global imaging.

If the transfer angle between two flybys is 360 degrees (that is, the two flybys occur with the same satellite at the same place), the orbit connecting the two flybys is called a "resonant orbit." The period of a Titan-resonant orbit is an integer multiple of Titan's orbital period. The plane of the transfer orbit between any two flybys is formed by the position vectors of the flybys from Saturn.

If the transfer angle is either 360 degrees (that is, the two flybys occur with the same satellite at the same place) or 180 degrees, an infinite number of orbital planes connects the flybys. In this case, the plane of the transfer orbit can be inclined significantly to the planet's equator. Any inclination can be chosen for the transfer orbit, as long as sufficient bending is available from the flyby to get to that inclination. If a spacecraft's orbital plane is significantly inclined to the equator, the transfer angle between any two flybys forming this orbital plane must be nearly 180 or 360 degrees.

If the angle between the position vectors is other than 180 or 360 degrees — as is usually the case — the orbital plane formed by the position vectors of the two flybys is unique and lies close to the satellites' orbital planes (except for Iapetus), which are close to Saturn's equator. In this case, the orbit is "nonresonant." Nonresonant orbits have orbital periods that are not integer multiples of Titan's period. Nonresonant Titan–Titan transfer orbits connect inbound Titan flybys to outbound Titan flybys, or vice versa.

General Tour Strategy

This section contains specific descriptions of the segments in the reference tour T18-3.

Titan 1–Titan 2. The first three Titan flybys reduce orbital period and inclination. The Orbiter's inclination is reduced to near zero with respect to Saturn's equator only after the third flyby; so, these three flybys must all take place at the same place in Titan's orbit. The period-reducing flybys were designed to be inbound, rather than outbound, to accomplish the additional goal of rotating the line of apsides counterclockwise. This moves the apoapsis toward the Sun line in order to provide time for observations of Saturn's atmosphere, and to allow Saturn occultations on subsequent orbits to occur at distances closer to Saturn.

Titan 3. After the inclination has been reduced to near Saturn's equator, a targeted inbound flyby of Enceladus is achieved on the fourth orbit on the way to an outbound flyby of Titan on orbit five on March 31, 2005. Changing from an inbound to an outbound Titan flyby here orients the line of nodes nearly normal to the Earth line. This minimizes the inclination required to achieve an occultation of Saturn, preparing for the series of near-equatorial Saturn and ring occultations that follows.

A Titan flyby occurring normal to the Earth line can be inbound or outbound (like any Titan flyby). For Titan flybys occurring nearly over the dawn terminator as in the reference tour, the spacecraft is closer to Saturn during the occultation if the Titan flyby is outbound than if it is inbound. The science return from the occultation is much greater if the spacecraft is close to Saturn than if it is far away.

In particular, the antenna "footprint" projected on the rings is smaller when the occultation occurs closer to Saturn, improving the spatial resolution of the "scattered" radio signal observations. This is an important influence on the design of the tour. Lowering inclination to the equator, switching from inbound to outbound Titan flybys and rotating the orbit counterclockwise near the start of the tour all help keep the spacecraft close to Saturn during the subsequent series of equatorial occultations.

Titan 4–Titan 7. Here, the minimum inclination required to achieve equatorial occultations is about 22 degrees. The two outbound flybys on March 31 and April 16, 2005, in-

MAIN CHARACTERISTICS OF TOUR SEGMENTS

Titan Flyby	Comments
T1–T2	Reduce period and inclination, target for Probe mission.
T3	Rotate counterclockwise and transfer from inbound to outbound.
T4–T7	Raise inclination for eight equatorial Saturn–ring occultations and lower again to equator.
T8–T15	Rotate clockwise toward anti-Sun direction.
T16	Increase inclination and rotate for magnetotail passage.
T17–T31	180-degree transfer sequence (including several revolutions for ring observations).
T32–T34	Target to close icy satellite flybys of Enceladus, Rhea, Dione and Iapetus.
T35–T43	Increase inclination to 71 degrees (maximum value in tour).

crease inclination to this value. The second of these also changes the period to 18.2 days. At this period, seven orbiter revolutions and eight Titan revolutions are completed before the next Titan flyby, producing seven near-equatorial occultations of Earth by Saturn (one on each orbit). On all eight of these revolutions, the Orbiter crosses Saturn's equator near Enceladus' orbit; on the fourth revolution, the second targeted flyby of Enceladus occurs. Enceladus' gravity is too weak to displace inclination significantly from the value required to achieve occultations. The Titan flybys on August 22 and September 7, 2005, reduce inclination once again to near Saturn's equator.

Titan 8–Titan 15. After inclination is reduced and the spacecraft's orbital plane again lies near Saturn's equator, a series of alternating outbound/period-reducing and inbound/period-increasing flybys — lasting about 10 months — is used to rotate the orbit clockwise toward the magnetotail. The first flyby in this series occurs on September 26, 2005, and the last occurs on June 1, 2006.

Titan 16. After rotating the orbit to place apoapsis near the anti-Sun line, inclination is raised to about 10 degrees with one flyby on July 6, 2006, to achieve passage through the current sheet in the magnetotail region. At distances this far from Saturn, the current sheet is assumed to be swept away from Saturn's equatorial plane by the solar wind. After this flyby, apoapsis distance is about 49 R_S, exceeding the 40 R_S magnetospheric and plasma science requirement associated with magnetotail passage. Besides providing a magnetotail passage, this inclination-raising flyby is the first of a sequence of flybys that makes up the 180-degree transfer sequence described next.

ON A FOUR-YEAR TOUR

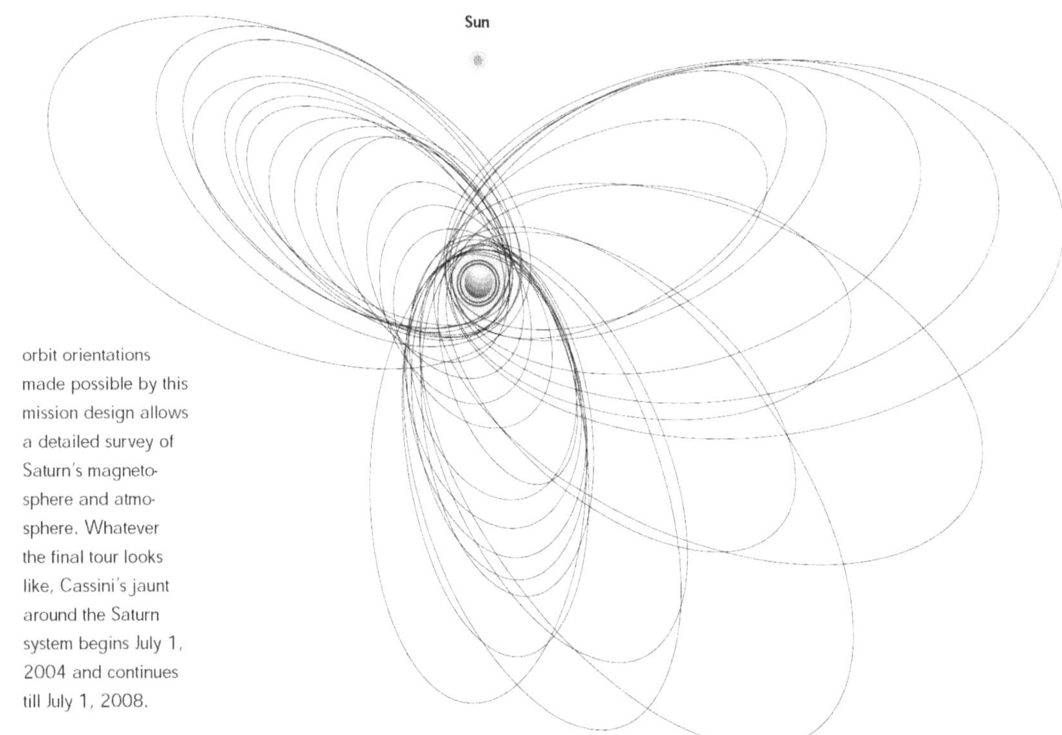

This view from above Saturn's north pole shows all possible orbits in a rotating coordinate system, in which the Sun direction is fixed, for a possible Saturn system tour referred to by mission designers as "Tour T18-3." This type of diagram is often referred to as a "petal" plot due to its resemblance to the petals of a flower. The broad range of orbit orientations made possible by this mission design allows a detailed survey of Saturn's magnetosphere and atmosphere. Whatever the final tour looks like, Cassini's jaunt around the Saturn system begins July 1, 2004 and continues till July 1, 2008.

Titan 17–Titan 31. This series of flybys completes a 180-degree transfer sequence. The first several flybys of this sequence — all inbound — are used to raise inclination as quickly as possible using the minimum altitude of 950 kilometers at each flyby. The Titan flyby of August 22, 2006, reduces the period to 16 days, as well as raising inclination. The period is then kept constant at 16 days as inclination is raised, except during an interval of 48 days between flybys on October 25 and December 12, 2006. The flyby on October 25, 2006, reduces the period to 12 days (a resonance of three Titan revs for every four spacecraft revs) in order to provide extra spacecraft revs between Titan flybys for observing the rings, for which the geometry is particularly favorable at this point in the tour. The flyby on December 12, 2006, increases the period back to 16 days.

As inclination is raised, the periapsis radius increases and apoapsis radius decreases until the orbit is nearly circularized at an inclination of about 60 degrees. The Orbiter's trajectory then crosses Titan's orbit at not one, but two points (the ascending and descending nodes), making possible a 180-degree transfer from an inbound Titan flyby to an outbound Titan flyby. After this 180-degree transfer is accomplished, the next seven Titan flybys, all of which are outbound, reduce inclination as quickly as possible to near Saturn's equator. This 180-degree transfer flyby sequence (raising inclination, accomplishing the 180-degree transfer, then lowering inclination again) rotates the line of apsides about 120 degrees so that apoapsis lies between the Sun line and Saturn's dusk terminator.

Titan 32–Titan 34. The flybys immediately following the completion of the 180-degree transfer sequence and the return of the spacecraft's orbital plane to near Saturn's equator are used to target flybys of Enceladus, Rhea, Dione and Iapetus. The Enceladus and Rhea flybys occur on successive orbits (46 and 47) between the Titan flybys on May 28 and July 18, 2007. The Titan flyby on September 1, 2007, raises inclination to seven degrees to target to Iapetus.

Titan 35–Titan 43. Following the Iapetus flyby on September 18, 2007, the Orbiter is targeted to an outbound Titan flyby on October 3, 2007, which places the line of nodes close to the Sun line. Starting with this flyby, the rest of the tour is devoted to a "maximum-inclination sequence" of flybys designed to raise inclination as high as possible for ring observations and in situ fields and particles measurements (in this case, to about 75 degrees). In this reference tour, the orbits during this maximum-inclination flyby sequence are oriented nearly toward the Sun, opposite the magnetotail, to ensure several occultations of Earth by Saturn and the rings at close range.

During this flyby sequence, first, orbit "cranking" and then, orbit "pumping" (after a moderate inclination has been achieved) are used to increase inclination, eventually reducing the orbit period to just over seven days (nine Orbiter revolutions, four Titan revolutions). The closest approach altitudes during this sequence are kept at the minimum allowed value to maximize gravitational assist at each flyby.

End of Mission

The reference tour ends on July 1, 2008, four years after insertion into orbit about Saturn, 33.5 days and four and a half spacecraft revs after the last Titan flyby (which occurs on May 28, 2008). The spacecraft's orbital period is 7.1 days (a resonance of four Titan revs to nine spacecraft revs), its inclination is 71.1 degrees and its periapsis radius is four R_S.

The aimpoint at the last flyby is chosen to target the orbiter to a Titan flyby on July 31, 2008 (64 days after the last flyby in the tour), providing the opportunity to proceed with more flybys during an extended mission, if resources allow. Nothing in the design of the tour precludes an extended mission.

CHAPTER 8

Vehicle of Discovery

The Cassini Orbiter, with its scientific instruments, is an amazing "tool" for making remote observations. However, much work has to be done to get the spacecraft to its destination, and that is the story of mission design and execution, covered in the previous chapter, and of mission operations, covered in Chapter 10. In this chapter, we will examine the structure and design of the Cassini spacecraft, including onboard computers, radio receivers and transmitters, data storage facilities, power supplies and other items to support data collection by the science payload. The instruments themselves, on the Orbiter and within the Huygens Probe, are covered in Chapter 9.

Orbiter Design

The functions of the Cassini Orbiter are to carry the Huygens Probe and the onboard science instruments to the Saturn system, serve as the platform from which the Probe is launched and science observations are made and store information and relay it back to Earth.

The design of the Orbiter was driven by a number of requirements and challenges that make Cassini–Huygens different from most other missions to the planets. Among these are the large distance from Saturn to the Sun and Earth, the length of the mission, the complexity and volume of the science observations and the spacecraft's path to Saturn. That path includes four "gravity assists" from planets along the way to Saturn.

General Configuration. The main body of the Orbiter is a stack of four main parts: the high-gain antenna (provided by the Italian space agency), the upper equipment module, the propulsion module and the lower equipment module. Attached to this stack are the remote-sensing pallet and the fields and particles pallet, both with their science instruments, and the Huygens Probe system (provided by the European Space Agency). The overall height of the assembled spacecraft is 6.8 meters, making Cassini–Huygens the largest planetary spacecraft ever launched.

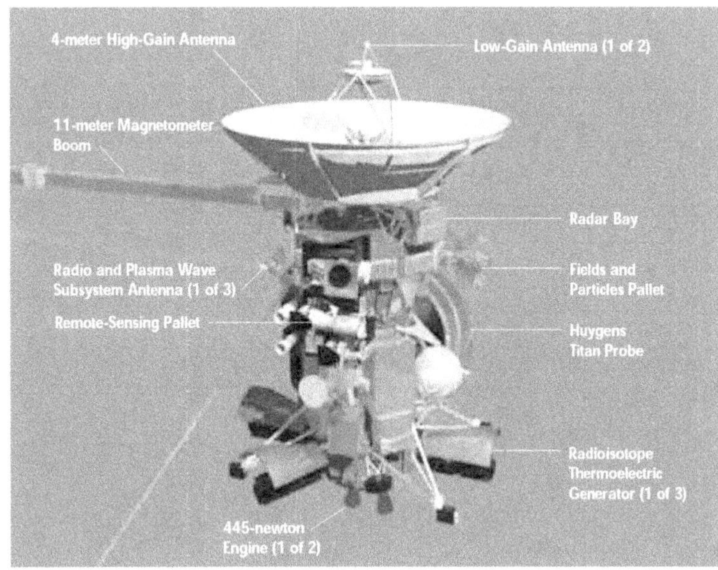

This illustration shows the main design features of the Cassini spacecraft, including the Huygens Titan Probe.

The Huygens Probe contains instruments for six investigations. The Orbiter carries instrumentation for 12 investigations. Four science instrument packages are mounted on the remote-sensing pallet; three more are mounted on the fields and particles pallet. Two are associated with the high-gain antenna (HGA). The magnetometer is mounted on its own 11-meter boom. The remaining instruments are mounted directly on the upper equipment module.

Power Source. Because of Saturn's distance from the Sun, solar panels of any reasonable size cannot provide sufficient power for the spacecraft. To generate enough power, solar arrays would have to be the size of a couple of tennis courts and would be far too heavy to launch.

The Cassini Orbiter will get its power from three radioisotope thermoelectric generators, or RTGs, which use heat from the natural decay of plutonium to generate direct current electricity. These RTGs are of the same design as those already on the Galileo and Ulysses spacecraft and have the ability to operate many years in space. At the end of the 11-year Cassini–Huygens mission, they will still be capable of producing 630 watts of power! The RTGs are mounted on the lower equipment module and are among the last pieces of equipment to be attached to the spacecraft prior to launch.

Fault Protection. The distance of Saturn from Earth is especially important, because it affects communication with the spacecraft. When Cassini is at Saturn, it will be between 8.2 and 10.2 astronomical units (AU) from Earth (one AU is the distance from Earth to the Sun, or 150 million kilometers). Because of this, it will take 68–84 minutes for signals to travel from Earth to the spacecraft, or vice versa.

In practical terms, this means that mission operations engineers on the ground cannot give the spacecraft "real-time" instructions, either for day-to-day operations or in case of unexpected events on the spacecraft. By the time ground personnel become aware of a problem and respond, nearly three hours will have passed. Onboard fault protection is therefore essential to the success of the spacecraft. The Cassini–Huygens spacecraft system fault protection is designed to ensure that the spacecraft can take care of itself in the event of onboard problems long enough to permit ground personnel to study the problem and take appropriate action. In practical terms, this means that if a fault is detected that might pose a substantial risk to any part of the spacecraft, onboard computers automatically initiate appropriate "safing" actions. These may include terminating preprogrammed activities and establishing a safe, commandable and relatively inactive spacecraft state for up to several weeks without ground intervention.

Command and Data Subsystem. The primary responsibility for command, control (including the fault protection

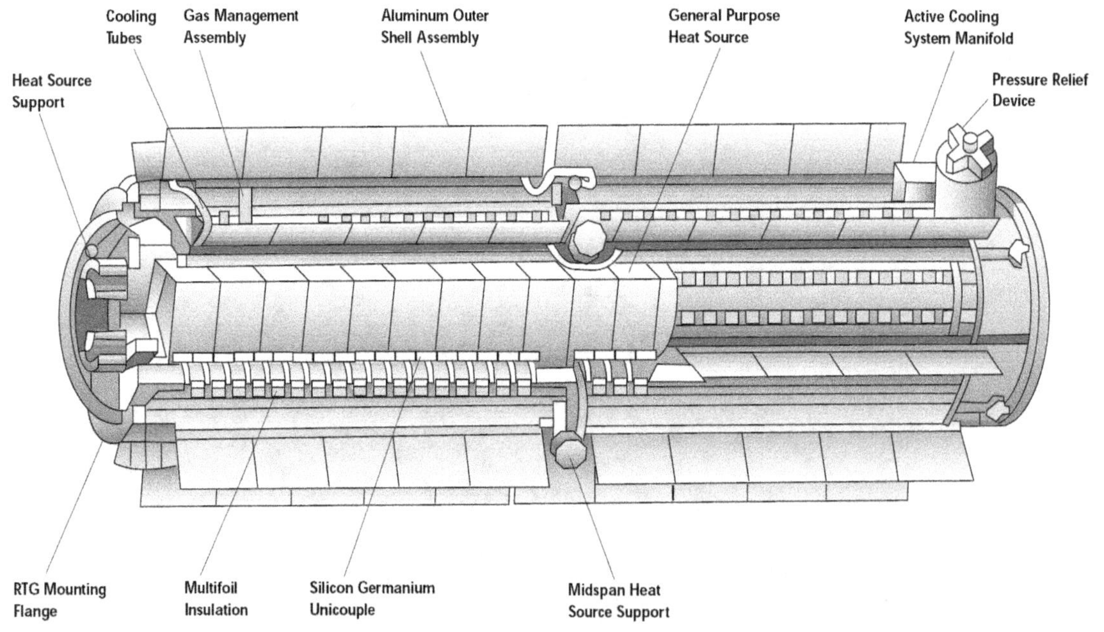

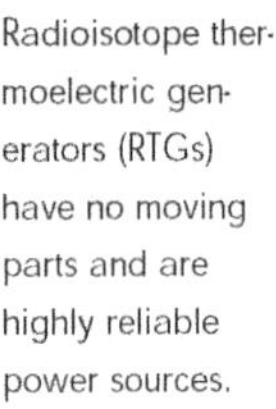

Radioisotope thermoelectric generators (RTGs) have no moving parts and are highly reliable power sources.

discussed earlier) and data handling is performed by the actively redundant command and data subsystem (CDS). This computer executes sequences of stored commands, either as a part of a normal preplanned flight activity or as a part of fault-protection routines. The CDS also processes and issues real-time commands from Earth, controls and selects data modes and collects and formats science and engineering data for transmission to Earth.

The CDS electronics are located in Bay 8 of the 12-bay upper equipment module. Commands and data from the CDS to each instrument and data from the instruments are handled by bus interface units (BIUs), located in the electronics boxes of each instrument. The BIUs are also used by the CDS to control data flow and allowable power states for each instrument. The CDS can accommodate data collection from the instruments and engineering subsystems at a combined rate in excess of 430,000 bits per second while still carrying on its command and control functions!

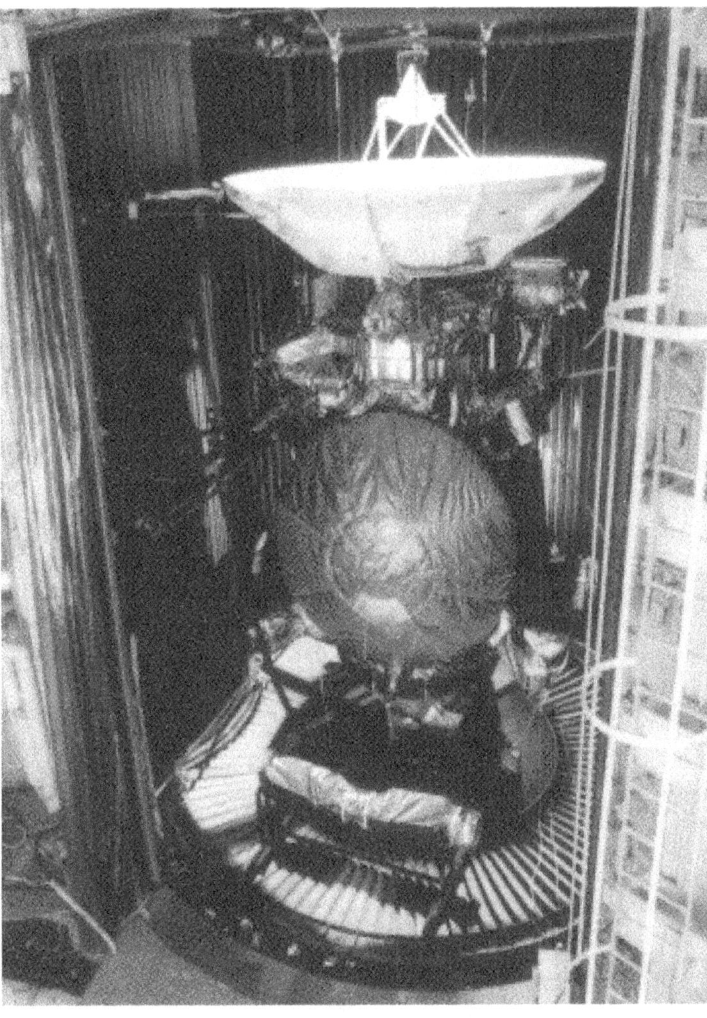

The Cassini–Huygens spacecraft, in launch configuration, in the Jet Propulsion Laboratory's High Bay.

Solid-State Recorders. The Cassini–Huygens spacecraft has been designed with a minimum of movable parts. In keeping with that design philosophy, data storage is accomplished by means of solid-state recorders rather than tape recorders. The two redundant solid-state recorders each had a storage capacity of two gigabits when they were built. Storage capacity is guaranteed to be at least 1.8 gigabits 15 years after launch. The solid-state recorders are located in Bay 9 of the upper equipment module.

The solid-state recorders have the ability to record and read out data simultaneously, record the same data simultaneously in two different locations on the same recorder and record simultaneously on both recorders. The recorders will be used to buffer essentially all of the collected data, permitting data transmission to Earth to occur at the highest available rates, rather than being restricted to instantaneous data collection rates. They can also be partitioned by command from the CDS. The recorders will be used to store backup versions of memory loads for almost all computers on the spacecraft and keep a running record of recent engineering activities to assist in the analysis of possible problems.

Attitude and Articulation Control. The attitude and articulation control subsystem (AACS) is primarily responsible for maintaining the orientation of

Cassini–Huygens in space. Specifically, the AACS is required to do the following tasks:

- Acquire a "fix" on the Sun following separation of the spacecraft from the launch vehicle.
- Point the antenna (either the high-gain or one of the two low-gain antennas) toward Earth when required.
- Point the high-gain antenna toward the Huygens Probe during its three-hour data collection period as it descends through the atmosphere and lands on Titan's surface.
- Point the high-gain antenna at appropriate radar or radio science targets.
- Point the instruments on the remote-sensing pallet toward targets that are themselves in motion relative to the spacecraft.
- Stabilize the spacecraft for Probe release and gravity wave measurements.
- Turn the spacecraft at a constant rate around the axis of the high-gain antenna for fields and particles measurements during transmission of data to Earth or receipt of commands from Earth.
- Point one of the two redundant main propulsion engines in the desired direction during main engine burns.
- Perform trajectory correction maneuvers of smaller magnitude using the onboard thrusters.
- Provide sufficient data in the transmitted engineering data to support science data interpretation and mission operations.

The two redundant computers for the AACS are located in Bays 1 and 10 of the upper equipment module, but parts of the AACS are spread across the spacecraft. Redundant Sun sensors are mounted to the high-gain antenna. Redundant stellar reference units are mounted on the remote sensing platform. Three mutually perpendicular reaction wheels are mounted on the lower equipment module: A fourth reaction wheel, mounted on the upper equipment module, is a backup that can be rotated to be parallel to any one of the three other reaction wheels. Redundant inertial reference units, consisting of four hemispherical resonator gyroscopes each, are mounted to the upper equipment module. The main engine actuators and electronics are mounted near the bottom of the propulsion module. An accelerometer, used to measure changes in the spacecraft's velocity, shares Bay 12 of the upper equipment module with the imaging science electronics.

A sophisticated pointing system known as inertial vector propagation is programmed into the AACS computers. It keeps track of spacecraft orientation, the direction and distance of the Sun, Earth, Saturn and other possible remote-sensing targets in the Saturn system and the spacecraft-relative pointing directions of all the science instruments — and thereby points any specified instrument at its selected target. The AACS uses the stellar reference unit to determine the orientation of the spacecraft by comparing stars seen in its 15-degree field of view to a list of more than 3000 stars stored in memory.

Propulsion Module Subsystem. The largest and most massive subsystem on the spacecraft is the propulsion module subsystem (PMS). It consists of a bipropellant element for trajectory and orbit changes and a hydrazine element for attitude control, small maneuvers and reaction wheel desaturation ("unloading").

The bipropellant components are monomethyl hydrazine (fuel) and nitrogen tetroxide (oxidizer), and are identical and stacked in tandem inside the cylindrical core structure, with the fuel closer to the spacecraft high-gain antenna. The fuel tank is loaded with 1130 kilograms of monomethyl hydrazine; the oxidizer tank contains 1870 kilograms of nitrogen tetroxide. Their combined mass at launch, 3000 kilograms, constitutes more than half the total mass

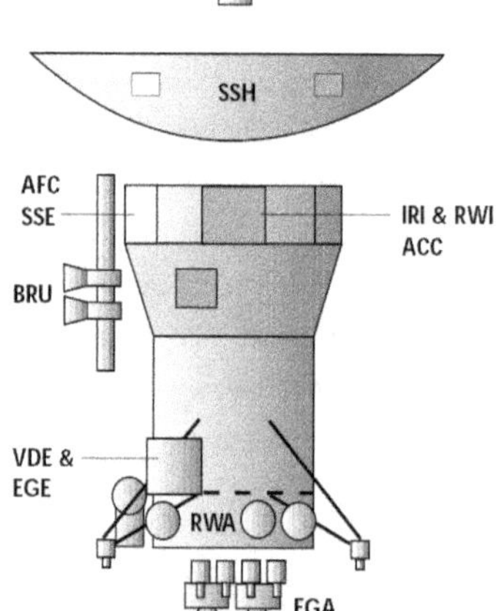

A diagram of the Cassini spacecraft showing elements of the propulsion module subsystem. (Acronyms are defined in Appendix B.)

of the spacecraft! Below the oxidizer tank are the redundant main engines, which can be articulated in each of two axes by the AACS. A heat shield protects the gimbal assembly and the spacecraft from engine heat.

The hydrazine tank is spherical and is mounted external to the PMS cylindrical structure. Hydrazine is used to power the small thrusters as commanded by the AACS. The 16 thrusters are located in clusters of four each on four struts extending outward from the bottom of the PMS. At launch, the tank holds 132 kilograms of hydrazine.

A larger cylindrical tank with rounded ends, also exterior to the PMS cylindrical structure but on the opposite side from the hydrazine tank, holds nine kilograms of helium. The helium tank supplies pressurant to expel propellants from the two bipropellant tanks and from the hydrazine tank.

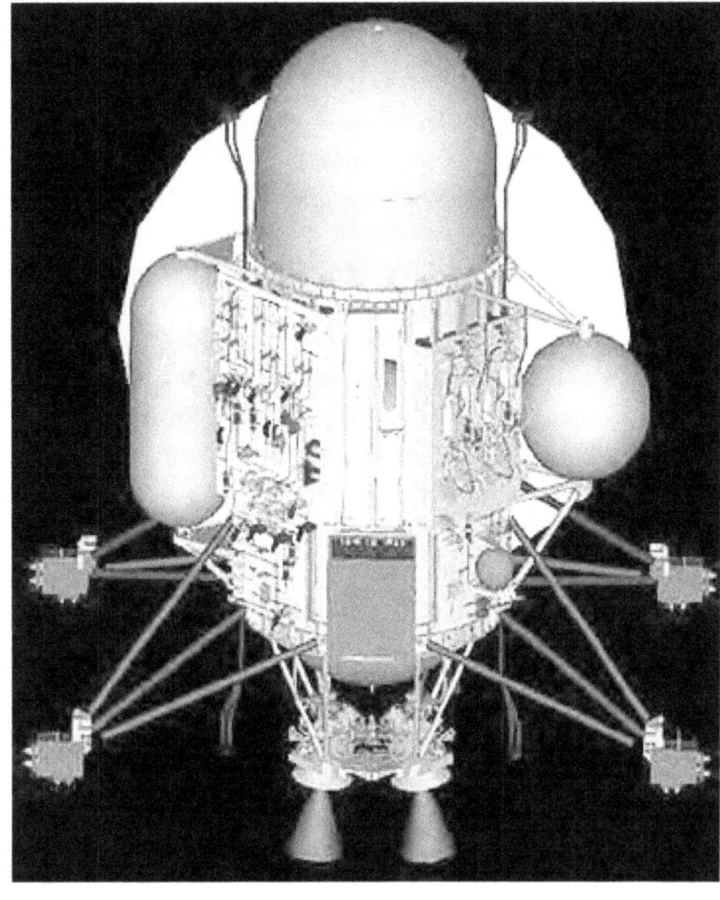

The propulsion module subsystem was tested in the Jet Propulsion Laboratory's Spacecraft Assembly Facility.

Telecommunications. The long-distance communications functions on the spacecraft are performed by the radio frequency subsystem (RFS) and the antenna subsystem. For telecommunications from the spacecraft to Earth, the RFS produces an X-band carrier at a frequency of 8.4 gigahertz, modulates it with data received from the CDS, amplifies the carrier and delivers the signal stream to the antenna subsystem for transmission. In the opposite direction, the RFS accepts X-band ground commands and data signals from the antenna subsystem at a frequency of 7.2 gigahertz, demodulates them and delivers the telemetry to the CDS for storage and/or execution. The RFS is contained in Bays 5 and 6 of the upper equipment module.

The antenna subsystem includes the four-meter-diameter high-gain antenna (HGA) and two low-gain antennas (LGA1 and LGA2). LGA1 is attached to the secondary reflector of the HGA and has a hemispherical field of view centered on the HGA field of view. LGA2 is mounted on the lower equipment module and has a hemispherical field of view centered on a direction approximately perpendicular to the HGA field of view. The two low-gain antennas are primarily used for communications during the first two and a half years after launch, when the HGA needs to be pointed at the Sun (to provide shade for most of the spacecraft subsystems) and cannot therefore be pointed at Earth. After the first two and a half years, the high-gain antenna is used almost exclusively for communications.

The spacecraft receives commands and data from Earth at a rate of 1000 bits per second during HGA operations. It transmits data to Earth at various rates, between 14,220

The Cassini high-gain antenna was provided by the Italian space agency.

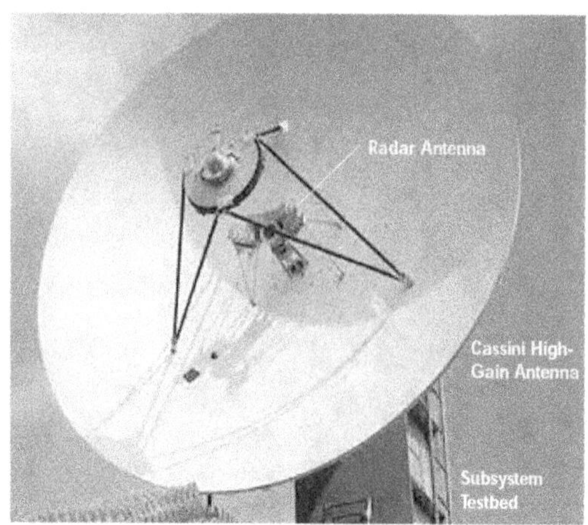

and 165,900 bits per second. Lower rates of receipt and transmission apply for low-gain antenna operations. In general, during operations from Saturn, data will be recorded on the solid-state recorders for 15 hours each day, while the HGA is not pointed at Earth. Then, for nine hours each day (generally during Goldstone, California, tracking station coverage) the data from the solid-state recorders will be played back while data collection from the fields and particles investigations continues. In this fashion, approximately one gigabit of data can be returned each "low-activity" day via a 34-meter Deep Space Network antenna. Similarly, using a 70-meter antenna on "high-activity" days, approximately four gigabits of data can be returned in nine hours.

In addition to redundant X-band receivers and transmitters, the HGA also houses feeds for a K_a-band receiver, a K_a-band transmitter and an S-band transmitter, all for radio science. Feeds for K_u-band transmitters and receivers for radar science are also on the HGA, as is a feed for an S-band receiver for receipt of the Huygens Probe data.

Power and Pyrotechnics Subsystem. The radioisotope thermoelectric generators (discussed earlier) are a part of the power and pyrotechnics subsystem (PPS). The RTGs generate the power used by the spacecraft; the power conditioning equipment conditions and distributes that power to the rest of the spacecraft; the pyro switching unit provides redundant power conditioning and energy storage and, upon command, redundant power switching for firing electro-explosive or pyro devices.

The power conditioning equipment converts the RTG output to provide a regulated 30-volt direct current power bus. It also provides the capability to turn power on and off to the various spacecraft power users in response to commands from the CDS. If any power user experiences an overcurrent condition, the power conditioning equipment detects that condition; if the level of overcurrent exceeds a predetermined level, the power to that user is switched off.

The pyro switching unit controls the firing of pyro devices on 32 commandable circuits, most of which open or shut valves to control the pressures and flows within the propulsion module plumbing. The pyro devices are also used to:

- Separate the spacecraft from the launch vehicle after launch.
- Pull the pin that unlatches the spare reaction wheel articulation mechanism.
- Pull the pin that permits the Radio and Plasma Wave Science instrument's Langmuir probe to deploy.
- Jettison science instrument covers.
- Separate the Huygens Probe from the Cassini Orbiter.
- Jettison the articulated cover that protects the main engine nozzles from damage by high-velocity particles in space (in the event the cover sticks in a closed position).

Temperature Control Subsystem. The Cassini spacecraft has many electrical and mechanical units that are sensitive to changes in temperature. The function of the temperature control subsystem (TCS) is to keep these units within their specified temperature limits while they are on the spacecraft, during both prelaunch and post-launch activities. The TCS monitors temperatures by means of electrical temperature sensors on all critical parts of the spacecraft. The TCS then controls the temperature of the various units by using one or more of the following techniques:

- Turning electrical heaters on or off to raise or lower temperatures.

- Opening or shutting thermal louvers (which resemble Venetian blinds) to cool or heat electronics in one of the bays of the upper equipment module.
- Using small radioisotope heater units to raise the temperatures of selected portions of the spacecraft that would otherwise cool to unacceptably low temperatures.
- Installing on the spacecraft thermal blankets (gold or black in color) to prevent heat from escaping to space.
- Installing thermal shields above the RTGs to keep them from radiating excessive heat to sensitive science instruments.
- Taking advantage of the orientation of the spacecraft to keep most of the spacecraft in the shadow of the high-gain antenna or the Huygens Probe.

From launch until Cassini–Huygens is well beyond the distance of Earth, the spacecraft is oriented to point the high-gain antenna at the Sun, thereby shading most of the rest of the spacecraft. Deviations from this orientation are permitted for a maximum of half an hour for each occurrence until the spacecraft reaches a distance of 2.7 AU (405 million kilometers) from the Sun. After that time, the heat input from the Sun will have diminished sufficiently to permit almost any spacecraft orientation.

Although they are not a part of the TCS responsibility, several instruments have radiator plates to cool their detectors. Even at the distance of Saturn from the Sun, spacecraft orientations that point those radiator plates in the same half of the sky as the Sun can severely degrade the data collected by some of the science instruments. In at least one case, it is even necessary to simultaneously avoid having Saturn illuminate the radiator plate(s).

New Technology. Whereas previous interplanetary spacecraft used onboard tape recorders, Cassini will pioneer a new solid-state data recorder with no moving parts. The recorder will be used in more than 20 other NASA and non-NASA missions.

Similarly, the main onboard computer, which directs Orbiter operations, uses a novel design incorporating new families of electronic chips. Among them are very high speed integrated circuit (VHSIC) chips. The computer also contains powerful new application-specific integrated circuit (ASIC) parts, each replacing a hundred or more traditional chips.

Also on the Orbiter, the power system will benefit from an innovative solid-state power switch that will eliminate the rapid fluctuations or "transients" that can occur with conventional power switches. This will significantly extend component lifetime.

Past interplanetary spacecraft have used massive mechanical gyroscopes to provide inertial reference during periods when Sun and star references were unavailable. Cassini employs hemispherical resonator gyroscopes that have no moving parts. These devices utilize tiny "wine glasses" about the size of a dime that outperform their more massive counterparts both in accuracy and in expected lifetime.

Huygens Probe Design

The Huygens Probe, supplied by the European Space Agency (ESA), will scrutinize the clouds, atmosphere and surface of Saturn's moon Titan. It is designed to enter and brake in

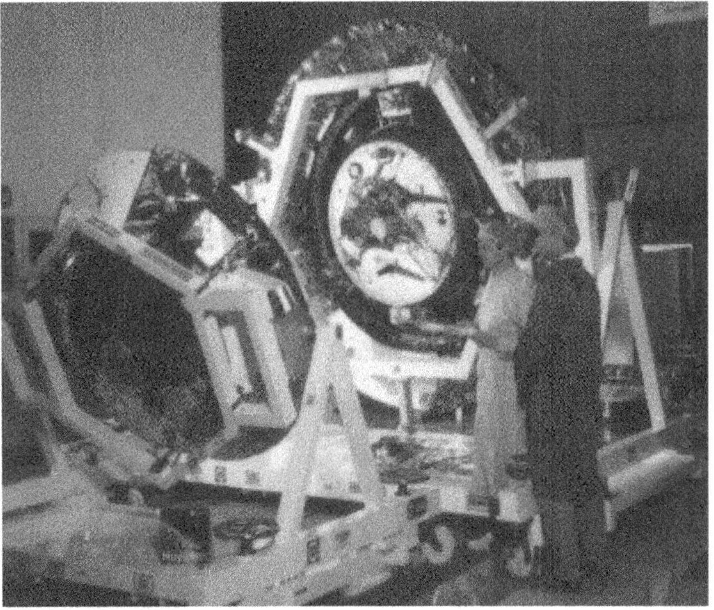

Final European Space Agency inspection of the Huygens Probe prior to shipment to the United States for integration with the Cassini Orbiter.

The Huygens Probe carries six science investigations to study the clouds, atmosphere and surface of Titan. Three separate parachutes are carried in the Probe's parachute compartment. The heat shield, front shield and back cover are jettisoned before data collection begins.

Titan's atmosphere and parachute a fully instrumented robotic laboratory down to the surface. The Huygens Probe system consists of the Probe itself, which will descend to Titan, and the Probe support equipment, which will remain attached to the orbiting spacecraft.

The Probe will remain dormant throughout the 6.7-year interplanetary cruise, except for health checks every six months. These checkouts will follow preprogrammed descent scenario sequences as closely as possible, and the results will be relayed to Earth for examination by system and payload experts.

Prior to the Probe's separation from the Orbiter, a final health check will be performed. The "coast" timer will be loaded with the precise time necessary to turn on the Probe systems (15 minutes before the encounter with Titan's atmosphere), after which the Probe will separate from the Orbiter and coast to Titan for 22 days with no systems active except for its wake-up timer.

The Probe system is made up of a number of engineering subsystems, some distributed between the Probe and the Probe support equipment on the Orbiter. The Huygens payload consists of a complement of six science instrument packages.

Probe Support Equipment. The Probe support equipment includes the electronics necessary to track the Probe, recover the data gathered during its descent and process and deliver the data to the Orbiter, from which the data will be transmitted or "downlinked" to the ground.

The Probe engineering subsystems are the entry subsystem, the inner structure subsystem, the thermal control subsystem, the electrical power subsystem, the command and data management subsystem and the Probe data relay subsystem.

Entry Subsystem. The entry subsystem functions only during the release of the Probe from the Orbiter and its subsequent entry into the Titan atmosphere. It consists of three main elements: the spin–eject device that propels the Probe away from the Orbiter; a front shield covered with special thermal protection material that protects the Probe from the heat generated during atmospheric entry; and an aft cover, also covered with thermal protection material, to reflect away heat from the wake of the Probe during entry.

The Probe will be targeted for a high-latitude landing site on the "day" side of Titan and released from the Orbiter on November 6, 2004. The spin–eject device will impart a relative velocity of about 0.3 meter per second and a spin rate (for stabilization) of about seven revolutions per minute. The Probe's encounter with Titan is planned for November 27, when it will enter the atmosphere at a velocity of 6.1 kilometers per second. The entry phase will last about three minutes, during which the Probe's velocity will decrease to about 400 meters per second.

Three parachutes will be used during the Probe's descent. When the on-board accelerometers detect a speed of Mach 1.5 near the end of the deceleration phase, the two-meter-diameter pilot chute will deploy, pulling off the aft cover. This will be followed immediately by deployment of the 8.3-meter-diameter main parachute. About 30 seconds after deployment of the main chute, the Probe's velocity will have dropped from Mach 1.5 to Mach 0.6, and the front heat shield will be released.

At this point, scientific measurements can begin. About 15 minutes later, the main chute will be released and a smaller, three-meter drogue chute will permit the Probe to reach Titan's surface, about two and a half hours after data taking starts. The Probe will hit the surface at a velocity of about seven meters per second and is expected to survive the impact. The Probe will continue to transmit data for an additional 30–60 minutes.

Inner Structure Subsystem. The inner structure of the Probe consists of two aluminum honeycomb platforms and an aluminum shell. It is linked to the front heat shield and the aft cover by fiberglass struts and pyrotechnically operated release mechanisms. The central equipment platform carries, on both its upper and lower surfaces, the boxes containing the electrical subsystems and the science experiments. The upper platform carries the parachutes (when stowed) and the antennas for communication with the Orbiter.

Thermal Control Subsystem. At different times during the mission, the Probe will be subjected to extreme thermal environments requiring a variety of passive controls to maintain the required temperature conditions. For instance, during the two Venus flybys, the solar heat input will be high. The Probe will get some protection from the shadow of the high-gain antenna, and when the antenna is off Sun-point for maneuvers or communication, the Probe will be protected by multilayer insulation that will burn off during the later atmospheric entry.

The Probe will be at its coldest just after it separates from the Orbiter. To ensure that none of the equipment

AQ60 tiles, similar to those used on the Space Shuttle, will help to dissipate heat during the Huygens Probe's descent through Titan's atmosphere.

SPACECRAFT SUBSYSTEMS MASS AND POWER

Subsystem or Component	Mass, kilograms	Power, watts	Comments on Power Usage
Engineering Subsystems			
Structure Subsystem	272.6	0.0	
Radio Frequency Subsystem	45.7	80.1	During downlinking of data
Power and Pyrotechnics Subsystem	216.0	39.1	Shortly after launch
Command and Data Subsystem	29.1	52.6	With both strings operating
Attitude and Articulation Control Subsystem	150.5	115.3	During spindown of spacecraft
(Engine Gimbal Actuator)		31.0	During main engine burns
Cabling Subsystem	135.1	15.1	Maximum calculated power loss
Propulsion Module Subsystem (dry)*	495.9	97.7	During main engine burns
Temperature Control Subsystem	76.6	117.8	During bipropellant warmup
		6.0	Temperature fluctuation allowance
		2.0	Radiation and aging allowance
		20.0	Operating margin allowance
Mechanical Devices Subsystem	87.7	0.0	
Packaging Subsystem	73.2	0.0	
Solid-State Recorders	31.5	16.4	Both recorders reading and writing
Antenna Subsystem	113.9	0.0	
Orbiter Radioisotope Heater Subsystem	3.8	0.0	
Science Instrument Purge Subsystem	3.1	0.0	
System Assembly Hardware Subsystem	21.7	0.0	
Total Engineering Mass*	1756.6		
Science Instruments			
Radio Frequency Instrument Subsystem	14.4	82.3	Both S-band and K_a-band operating
Dual Technique Magnetometer	8.8	12.4	Both scalar and vector operations
Science Calibration Subsystem	2.2	44.0	In magnetometer calibration mode
Imaging Science (narrow-angle camera)	30.6	28.6	Active and operating
Imaging Science (wide-angle camera)	25.9	30.7	Active and operating
Visible and Infrared Mapping Spectrometer	37.1	24.6	In imaging mode
Radio and Plasma Wave Science	37.7	17.5	During wideband operations
Ion and Neutral Mass Spectrometer	10.3	26.6	Neutral Mass Spectrometer operating
Magnetospheric Imaging Instrument	29.0	23.4	High-power operations
Cosmic Dust Analyzer	16.8	19.3	Operating with articulation
Cassini Radar	43.3	108.4	Operating in imaging mode
Cassini Plasma Spectrometer	23.8	19.2	Operating with articulation
Ultraviolet Imaging Spectrograph	15.5	14.6	On in sleep state
Composite Infrared Spectrometer	43.0	43.3	Active and operating
Total Science Mass	338.2		
Huygens Probe (including Probe support)	350.0	249.8	During Probe checkouts
Launch Vehicle Adaptor Mass	136.0	0.0	
TOTALS*	2580.7	680.5	Power available at middle of tour

** Note that the masses given above do not include approximately 3141 kilograms of propellant and pressurant.*

falls below its storage temperature limits, the Probe will carry a number of radioisotope heater units that each generate one watt.

At described above under the entry subsystem, the front heat shield will protect the Probe during initial atmospheric entry. The front shield is covered with Space Shuttle–like tiles made of a material known as AQ60, developed in France. This material is essentially a low-density "mat" of silica fibers. The tile thickness of the front shield is calculated to ensure that the temperature of the structure will not exceed 150 degrees Celsius, below the melting temperature of lead. The rear side of the Probe will reach much lower temperatures, so a sprayed-on layer of "Prosial" silica foam material will be used on the rear shield. The overall mass of the thermal control subsystem will be more than 100 kilograms, or almost one-third of the entire Probe mass.

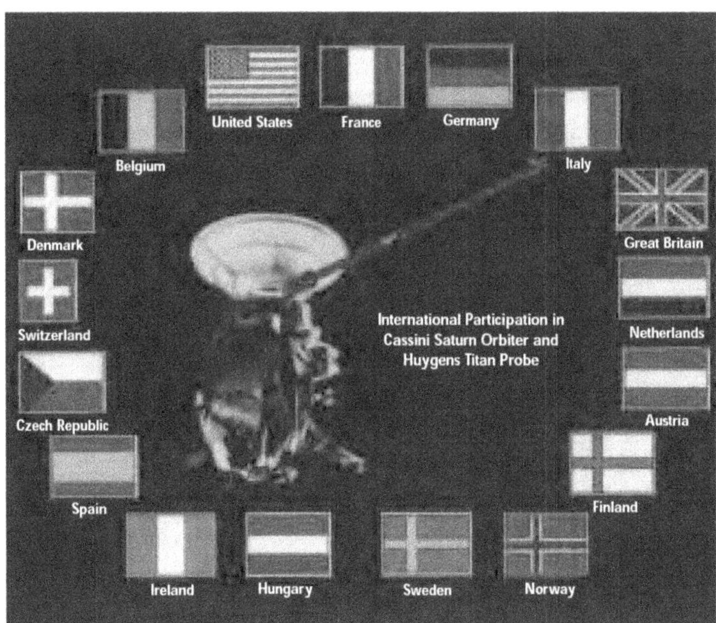

The Cassini–Huygens mission: an international collaboration.

Electrical Power Subsystem. During Probe checkout activities, the Probe will obtain power from the Orbiter via the umbilical cable. After separation, the Orbiter will continue to supply power to the Probe support equipment, but power for the Probe itself will be provided by five lithium–sulphur dioxide batteries.

Much of the battery power will be used to power the timer for the 22 days of "coasting" to Titan. The higher current needed for Probe mission operations is only required for the science data collection period, three to three and a half hours. The electrical power subsystem is designed to survive the loss of one of its five batteries and still support a complete Probe mission.

Command and Data Management. The command and data management subsystem provides monitoring and control of all Probe subsystem and science instrument activities. Specifically, this subsystem performs the following functions:

- Time the 22-day coast phase to Titan and switch the Probe on just prior to atmospheric entry.
- Control the activation of deployment mechanisms during the descent to Titan's surface.
- Distribute telecommands to the engineering subsystems and science instruments.
- Distribute to the science instruments a descent data broadcast providing a timeline of conditions on which the instruments can base the scheduling of mode changes and other operations.
- Collect science and housekeeping data and forward the data to the Orbiter via the umbilical cable (before Probe separation) or via the Probe data relay subsystem (during descent).

Probe Data Relay Subsystem. The Probe data relay subsystem provides the one-way Probe-to-Orbiter communications link and includes equipment on both Probe and Orbiter. The elements that are part of the Probe support equipment on the Orbiter are the Probe system avionics and the radio frequency electronics, including an ultrastable oscillator and a low-noise amplifier.

The Probe carries two redundant S-band transmitters, each with its own antenna. The telemetry in one link is delayed by about four seconds with respect to the other link, to prevent data loss if there are brief transmission outages. Reacquisition of the Probe's signal by the Orbiter would normally occur within this interval.

Concluding Remarks

The design of the Cassini–Huygens spacecraft is the end result of extensive trade-off studies that considered cost, mass, reliability, durability, suitability and availability of hardware.

To forestall the possibility of mechanical failures, moving parts were eliminated from the spacecraft wherever feasible. Early designs that included articulating instrument scan platforms or turntables were discarded in favor of body-fixed instruments whose pointing will require rotation of the entire spacecraft.

Tape recorders were replaced with solid-state recorders. Mechanical gyroscopes were replaced with hemispherical resonator gyroscopes. An articulated Probe relay antenna was discarded in favor of using the high-gain antenna to capture the Probe's signal.

Spacecraft engineers, both those who designed and built the hardware and those who will operate the spacecraft, relied heavily on extensive experience to provide a spacecraft design that is more sophisticated, reliable and capable than any other spacecraft ever built for planetary exploration. Because of that care in design, the Cassini–Huygens spacecraft will be far easier to operate and will return more science data about its targets than any prior planetary mission!

CHAPTER 9

Tools of Discovery

The 18 scientific instruments carried by the Cassini–Huygens spacecraft have been designed to perform the most detailed studies ever of the Saturn system. Twelve instruments are mounted on the Cassini Orbiter and six are carried inside the Huygens Probe. Each instrument will make its own unique measurements, providing comprehensive, synergistic information on Saturn, its rings, its satellites and its magnetosphere. Cassini–Huygens' scientific investigations will build on the wealth of data provided by Pioneer 11 and Voyager 1 and 2.

Orbiter Instruments Overview

The Cassini–Huygens mission is a cooperative, international endeavor. In addition to the United States' involvement, a number of other countries have provided instrument hardware and software as well as investigators to the instrument teams. The United States' international partners in one or more elements of the Cassini–Huygens mission are Austria, Belgium, the Czech Republic, Denmark, the European Space Agency, Finland, France, Germany, Hungary, Ireland, Italy, The Netherlands, Norway, Spain, Sweden, Switzerland and the United Kingdom. There are 12 scientific instruments aboard the Cassini Orbiter, which will spend four years studying the Saturn system in detail.

The Cassini–Huygens spacecraft stands almost seven meters tall and over four meters wide. In launch-ready configuration, the spacecraft weighs over 5600 kilograms.

All 12 Orbiter instruments are body-fixed to the spacecraft; some have articulating platforms that allow them to scan a portion of the sky without having to move the spacecraft. Each instrument has one or more microprocessors for internal control and data handling. The Orbiter experiments can be divided into three basic categories — optical remote sensing, microwave remote sensing and fields, particles and waves — and comprise a total of 27 sensors. The total mass of the science payload is 365 kilograms.

Orbiter Instrument Descriptions

Optical Remote Sensing

Imaging Science Subsystem. The optical remote-sensing Imaging Science Subsystem (ISS) will photograph a wide variety of targets — Saturn, the rings, Titan, the icy satellites and star fields — from a broad range of observing distances for various scientific purposes. General science objectives for the ISS include studying the atmospheres of Saturn and Titan, the rings of Saturn and their interactions with the planet's satellites and the surface characteristics of the satellites, including Titan. The following tasks are involved:

- Map the three-dimensional structure and motions in the Saturn and Titan atmospheres.

- Study the composition, distribution and physical properties of clouds and aerosols.
- Investigate scattering, absorption and solar heating in the Saturn and Titan atmospheres.
- Search for evidence of lightning, aurorae, airglow and planetary oscillations.
- Study gravitational interactions among the rings and the satellites.
- Determine the rate and nature of energy and momentum transfer within the rings.
- Determine ring thickness and the size, composition and physical nature of ring particles.
- Map the surfaces of the satellites, including Titan, to study their geological histories.
- Determine the nature and composition of the icy satellites' surface materials.
- Determine the rotation states of the icy satellites.

The ISS comprises a narrow-angle camera and a wide-angle camera. The narrow-angle camera provides high-resolution images of the target of interest, while the wide-angle camera provides extended spatial coverage at lower resolution. The cameras can also obtain optical navigation frames, which are used to keep the spacecraft on the correct trajectory.

OPTICAL REMOTE SENSING

Four optical remote-sensing instruments are mounted on a fixed, remote-sensing pallet, with their optical axes coaligned. The entire spacecraft must be rotated to point these instruments at the target of interest. The instruments include an Imaging Science Subsystem (ISS) comprising narrow- and wide-angle cameras, a Visible and Infrared Mapping Spectrometer (VIMS), a Composite Infrared Spectrometer (CIRS) and an Ultraviolet Imaging Spectrograph (UVIS). A fifth remote-sensing instrument, the imaging portion of the Magnetospheric Imaging Instrument (MIMI), is mounted on the upper equipment module, not far from the remote-sensing platform, and is also boresighted with the optical remote-sensing instruments. This experiment has the capability to image the charged particle population of Saturn's magnetosphere.

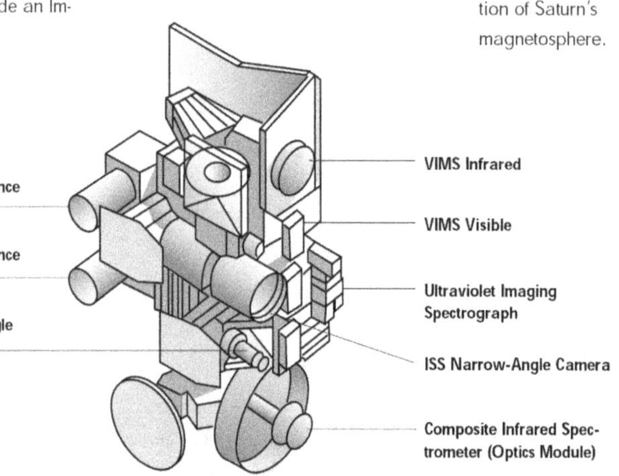

The Cassini–Huygens imagers differ primarily in the design of the optics. The wide-angle camera has refractive optics with a focal length of 200 millimeters, a focal length–diameter ratio (f number) of 3.5 and a 3.5-degree-square field of view (FOV). Refractive rather than reflective optics were chosen primarily to meet mass and cost constraints; these optics were available in the form of spares from the Voyager mission.

The narrow-angle camera has Ritchey–Chretien reflective optics with a 2000-millimeter focal length, an f number of 10.5 and a 0.35-degree FOV.

Filters are mounted in two rotatable wheels per camera; they have a two-wheel filter-changing mechanism whose design is derived from that of the Hubble Space Telescope's Wide Field and Planetary Camera 2. The wide-angle camera has 18 filters covering the range 380–1100 nanometers; the narrow-angle camera has 24 filters covering the range 200–1100 nanometers.

Shutters of the same type used on both Voyager and Galileo missions control exposure times: The shortest planned is five milliseconds and the longest is 20 minutes. The sensing element of each camera is a 1024 × 1024-element, solid-state array, or charge-coupled device (CCD). The CCD is phosphor-coated for ultraviolet response and radiator-cooled to 180 kelvins to reduce dark current (residual current in the CCD beyond that released by incident light).

Visible and Infrared Mapping Spectrometer. The Visible and Infrared Mapping Spectrometer (VIMS) will

The Imaging Science Subsystem (ISS) narrow-angle camera.

The Imaging Science Subsystem (ISS) wide-angle camera.

The Visible and Infrared Mapping Spectrometer (VIMS).

map the surface spatial distribution of the mineral and chemical features of a number of targets, including the rings and satellite surfaces and the atmospheres of Saturn and Titan. The VIMS science objectives are as follows:

- Map the temporal behavior of winds, eddies and other features on Saturn and Titan.
- Study the composition and distribution of atmospheric and cloud species on Saturn and Titan.
- Determine the composition and distribution of surface materials on the icy satellites.
- Determine the temperatures, internal structure and rotation of Saturn's deep atmosphere.
- Study the structure and composition of Saturn's rings.
- Search for lightning on Saturn and Titan and active volcanism on Titan.
- Observe Titan's surface.

The VIMS comprises a pair of imaging–grating spectrometers that are designed to measure reflected and emitted radiation from atmospheres, rings and surfaces to determine their compositions, temperatures and structures. The VIMS is an optical instrument that splits the light received from objects into its component wavelengths. The instrument uses a diffraction grating for this purpose.

The VIMS obtains information over 352 contiguous wavelengths from 0.35 to 5.1 micrometers. The instrument measures the intensities of individual wavelengths. The data are used to infer the composition and other properties of the object that emitted the light (such as a distant

The bar charts at right show the operating wavelength coverage for the optical remote-sensing investigations (above) and energy range for the fields, particles and waves investigations (below).

star), that absorbed specific wavelengths of the light as it passed through (such as a planetary atmosphere), or that reflected the light (such as a satellite surface). The VIMS provides images in which every pixel contains high-resolution spectra of the corresponding spot on the ground.

The VIMS has separate infrared and visible sensor channels. The infrared channel covers the wavelength range 0.85–5.1 micrometers. Its optics include an $f/3.5$ Ritchey–Chretien telescope that has an aperture of 230 millimeters, and a secondary mirror that scans on two axes to produce images varying in size from 0.03 degree to several degrees.

Radiation gathered by the telescope passes through a diffraction grating, which disperses it in wavelength. A camera refocuses it on to a linear detector array. The detector array is radiator-cooled to as low as 56 kelvins; the spectrometer's operating temperature is 125 kelvins.

The visible channel produces multispectral images spanning the spectral range 0.35–1.05 micrometers. It utilizes a Shafer telescope and a grating spectrometer. The silicon CCD array is radiator-cooled to 190 kelvins.

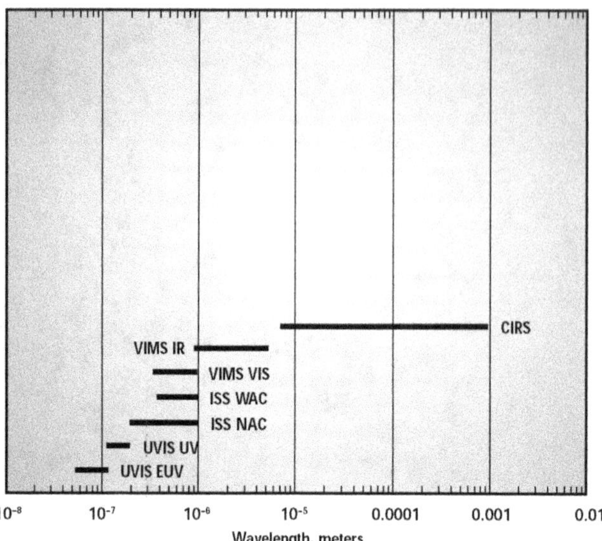

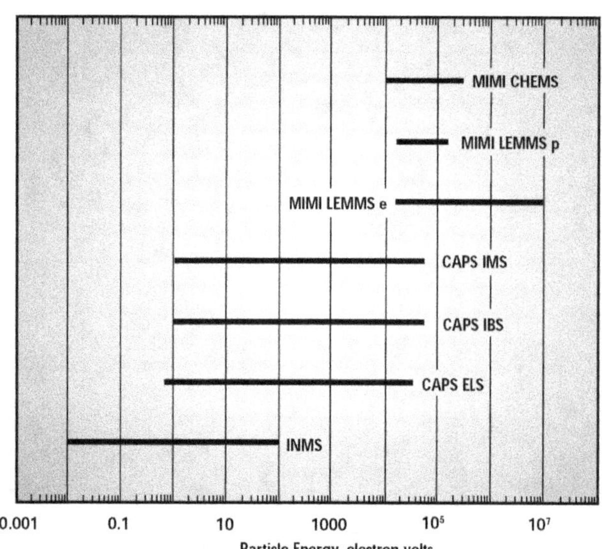

Composite Infrared Spectrometer. The Composite Infrared Spectrometer (CIRS) will measure infrared emissions from atmospheres, rings and surfaces. The CIRS will retrieve vertical profiles of temperature and gas composition for the atmospheres of Titan and Saturn, from deep in their tropospheres (lower atmospheres), to high in their stratospheres (upper atmospheres). The CIRS instrument will also gather information on the thermal properties and compositions of Saturn's rings and icy satellites.

The CIRS science objectives are as follows:

- Map the global temperature structure in the Saturn and Titan atmospheres.
- Map the global gas composition in the Saturn and Titan atmospheres.
- Map global information on hazes and clouds in the Saturn and Titan atmospheres.
- Collect information on energetic processes in the Saturn and Titan atmospheres.
- Search for new molecular species in the Saturn and Titan atmospheres.
- Map the global surface temperatures at Titan's surface.

• Map the composition and thermal characteristics of Saturn's rings and icy satellites.

The CIRS is a coordinated set of three interferometers designed to measure infrared emissions from atmospheres, rings and surfaces over wavelengths from 7 to 1000 micrometers. The CIRS uses a beamsplitter to divide incoming infrared light into two paths. The beamsplitter reflects half the energy toward a moving mirror and transmits half to a fixed mirror. The light is recombined at the detector. As the mirror moves, different wavelengths of light alternately cancel and reinforce each other at a rate that depends on their wavelengths. This information can be used to construct an infrared spectrum.

The CIRS data will be collected by two of the instrument's three interferometers, which are designed to make precise measurements of wavelength within a specific range of the electromagnetic spectrum. The CIRS far infrared interferometer covers the spectral range 17–1000 micrometers; its FOV is circular and 0.25 degree in diameter.

The CIRS mid infrared interferometer is a conventional Michelson instrument that covers the spectral range of 7–17 micrometers. The third CIRS interferometer is used for internal reference. Light enters the instrument through a 51-centimeter-diameter telescope and is sent to the interferometers. The mid infrared detectors consist of two 1 × 10 linear arrays. Each square detector in the arrays is 0.015 degree on a side. The mid infrared arrays are cooled to 70 kelvins by a passive radiator. The remainder of the instrument, including the infrared detectors, is cooled to 170 kelvins.

Ultraviolet Imaging Spectrograph. The Ultraviolet Imaging Spectrograph (UVIS) is a set of detectors designed to measure ultraviolet light reflected by or emitted from atmospheres, rings and surfaces to determine their compositions, distributions, aerosol content and temperatures. The UVIS will measure the fluctuations of sunlight and starlight as the Sun and stars move behind the rings and atmospheres of Saturn and Titan, and will determine the atmospheric concentrations of hydrogen and deuterium.

The UVIS science objectives involve the following:

- Map the vertical and horizontal compositions of the Saturn and Titan upper atmospheres.
- Determine the atmospheric chemistry occurring in the Saturn and Titan atmospheres.
- Map the distributions and properties of aerosols in the Saturn and Titan atmospheres.
- Infer the nature and characteristics of circulation in the Saturn and Titan atmospheres.
- Map the distributions of neutrals and ions in Saturn's magnetosphere.

The Composite Infrared Spectrometer (CIRS).

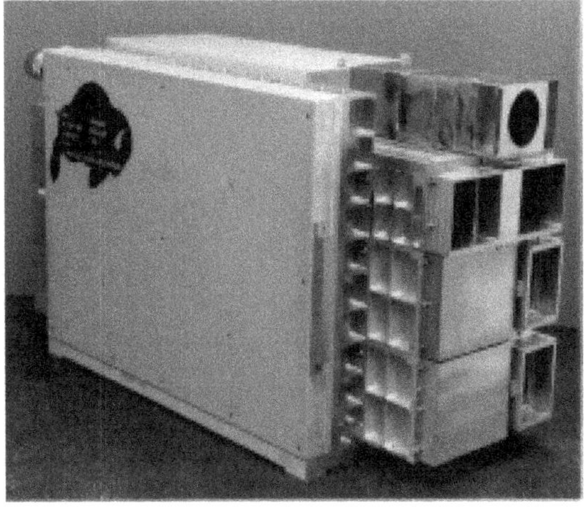

The Ultraviolet Imaging Spectrograph (UVIS).

- Study the radial structure of Saturn's rings by means of stellar occultations.
- Study surface ices and any tenuous atmospheres associated with the icy satellites.

The UVIS instrument is a two-channel imaging spectrograph (far and extreme ultraviolet) that includes a hydrogen–deuterium absorption cell and a high-speed photometer. An imaging spectrograph is an instrument that records spectral intensity information in one or more wavelengths of energy and then outputs digital data that can be displayed in a visual form, such as a false-color image.

Each of the two spectrographic channels utilizes a reflecting telescope, a grating spectrometer and an imaging, pulse-counting detector. The telescopes have a focal length of 100 millimeters and an FOV of 3.67 × 0.34 degrees. The far ultraviolet channel has a wavelength range of 115–190 nanometers; the range of the extreme ultraviolet channel is 55–115 nanometers. Spectral resolution ranges from 0.21 to 0.24 nanometers. For solar occultation observations, the extreme ultraviolet channel includes a mechanism that allows sunlight to enter when the Sun is 20 degrees off the telescope axis.

The absorption cell channel is a photometer that measures hydrogen and deuterium concentrations. The high-speed photometer measures undispersed light from its own parabolic mirror with a photomultiplier tube detector. The wavelength range for this photometer channel is 115–185 nanometers. The FOV is 0.34 × 0.34 degrees. The time resolution is 2 milliseconds.

Microwave Remote Sensing

Cassini Radar. The Cassini Radar (RADAR) will investigate the surface of Saturn's largest moon, Titan. Titan's surface is covered by a thick, cloudy atmosphere that is hidden to normal optical view, but can be penetrated by radar.

The RADAR uses the five-beam K_u-band (13.78 gigahertz) antenna feed on the high-gain antenna (HGA) to send RADAR transmissions toward targets. These signals, after reflection from the target, will be captured by the HGA and detected by the RADAR. The RADAR will also operate in a passive mode wherein the instrument will measure the blackbody radiation emitted by Titan. Science objectives of the RADAR include the following:

- Determine if large bodies of liquid exist on Titan, and if so, determine their distribution.
- Investigate the geological features and topography of Titan's solid surface.
- Acquire data on other targets (rings and icy satellites) as conditions permit.

The RADAR will take four types of observations: imaging, altimetry, backscatter and radiometry. In imaging synthetic aperture radar (SAR) mode, the instrument will record echoes of microwave energy off the surface of Titan from different incidence angles. The recorded echoes will allow construction of visual images of the target surface with a surface resolution of 0.35–1.7 kilometers.

Radar altimetry similarly involves bouncing microwave pulses off the subsatellite surface of the target body and measuring the time it takes the echo to return to the spacecraft. In this case, however, the goal will not be to create visual images but rather to obtain numerical data on the pre-

MICROWAVE REMOTE SENSING

Two microwave remote-sensing experiments, the Cassini Radar (RADAR) and the Radio Science Instrument (RSS), share the spacecraft's high-gain antenna (HGA). This antenna receives spacecraft commands from Earth and sends science and spacecraft data back to Earth; it is also used for communications with the Huygens Probe. Some RADAR components are housed in equipment bays in the upper equipment module, below the HGA; one piece, the radio frequency electronics subsystem (RFES), sits in a penthouse-like attachment between the upper equipment module and the HGA. The RSS components are housed in the spacecraft bays; the instrument also includes radio receiver elements located at NASA's Deep Space Network stations on Earth.

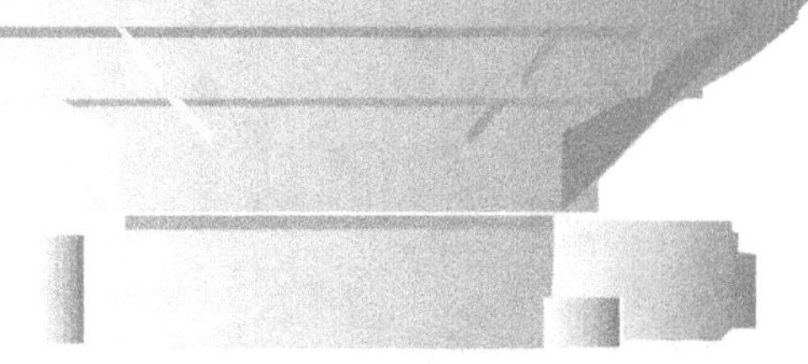

cise distance between the surface features of Titan and the spacecraft.

When these data are combined with the spacecraft ephemeris, a Titan topography map (altitude of surface features) can be created. The altimeter resolution is 24–27 kilometers horizontal, 90–150 meters vertical.

In scatterometry mode, the RADAR will bounce pulses off Titan's surface and then measure the intensity of the energy returning. This returning energy, or backscatter, is always less than the original pulse since surface features inevitably reflect the pulse in more than one direction. From the variations in these backscatter measurements, scientists can infer the composition and roughness of the surface.

Finally, in radiometry mode, the RADAR will operate as a passive instrument, simply recording the energy emanating from the surface of Titan. This information will tell scientists about Titan's surface properties. In the interferometry and scatterometry modes, the low-resolution (0.35-degree) data will be acquired from large amounts of Titan's surface by scanning the HGA.

The RADAR can operate at altitudes below approximately 100,000 kilometers. From 100,000 to 25,000 kilometers, the RADAR will operate in radiometry mode only to gather data about surface temperature and emissivity that is related to composition.

Between 22,500 and 9000 kilometers, the RADAR will switch between scatterometry and radiometry to obtain low-resolution global maps of Titan's surface roughness, backscatter intensity and thermal emissions. At altitudes between 9000 and 4000 kilometers, the instrument will switch between altimetry and radiometry, collecting surface altitude and thermal emission measurements along the suborbital track. Below 4000 kilometers, the RADAR will switch between imaging and radiometry.

At the RADAR's lowest planned altitude, 950 kilometers, the maximum image resolution will be 540 × 350 meters. In imaging mode, the RADAR will use all five beams in a fan shape to broaden the swath to approximately six degrees.

Radio Science Instrument. The Radio Science Instrument (RSS) will use the spacecraft radio and ground antennas — such as those of NASA's Deep Space Network (DSN) — to study the compositions, pressures and temperatures of the atmospheres and ionospheres of Saturn and Titan; the radial structure of Saturn's rings and the particle size distribution within the rings; and planet–satellite masses, gravity fields and ephemerides within the Saturn system. During the long interplanetary cruise to Saturn, the RSS will also be used — near solar oppositions to search for gravitational waves coming from beyond our solar system, and near solar conjunctions to test general relativity and study the solar corona. The instrument's science objectives include the following:

- Search for and characterize gravitational waves coming from beyond the solar system.
- Study the solar corona and general relativity when Cassini passes behind the Sun.
- Improve estimates of the masses and ephemerides of Saturn and its satellites.
- Study the radial structure of and particle size distribution in Saturn's rings.
- Determine temperature and composition profiles within the Saturn and Titan atmospheres.
- Determine temperatures and electron densities in the Saturn and Titan ionospheres.

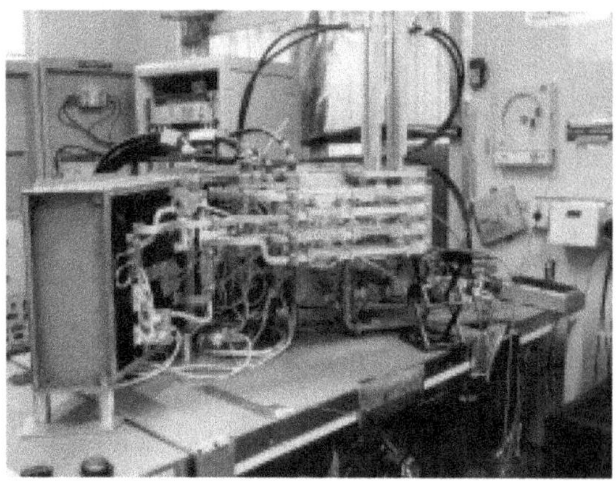

The Cassini Radar (top) and the Radio Science Instrument (RSS).

The RSS instrument is unique in that it consists of elements on the ground as well as on the spacecraft (see the sidebar on the next page). The ground elements are DSN 70-meter and 34-meter high-efficiency and beam waveguide stations.

On the spacecraft, the elements of the RSS are contained in the radio frequency subsystem and in the radio frequency instrument subsystem. A body-fixed HGA with its boresight along the spacecraft's *z*-axis will be pointed by moving the spacecraft. The HGA will be configured to operate at S-band, X-band and K_a-band to support radio science.

Occultation experiments measure the refractions, Doppler shifts (frequency shifts) and other modifications to radio signals that occur when the spacecraft is "occulted" by (passes behind) planets, moons, atmospheres or physical features such as planetary rings. From these measurements, scientists can derive information about the structures and compositions of the occulting bodies, atmospheres and the rings.

Cruise experiments and gravity field measurements use Doppler and ranging two-way measurements. The signal is generated from a reference on the ground, transformed to the transmitting frequencies, amplified and radiated through a Deep Space Network antenna. The spacecraft telecommunications subsystem receives the carrier signal, transforms it to downlink frequencies coherent with the uplink, amplifies it and returns it to Earth. The signal is detected through a ground antenna, amplified and saved for later analysis.

Fields, Particles and Waves

Cassini Plasma Spectrometer. The Cassini Plasma Spectrometer (CAPS) will measure the composition, density, flow velocity and temperature of ions and electrons in Saturn's magnetosphere. The CAPS science objectives are as follows:

- Measure the composition of ionized molecules originating from Saturn's ionosphere and Titan.
- Investigate the sources and sinks of ionospheric plasma — ion inflow/outflow and particle precipitation.
- Study the effect of magnetospheric and ionospheric interaction on ionospheric flows.
- Investigate auroral phenomena and Saturn kilometric radiation (SKR) generation.
- Determine the configuration of Saturn's magnetic field.
- Investigate plasma domains and internal boundaries.
- Investigate the interaction of Saturn's magnetosphere with the solar wind and solar-wind driven dynamics within the magnetosphere.
- Study the microphysics of the bow shock and magnetosheath.
- Investigate rotationally driven dynamics, plasma input from the satellites and rings and radial transport and angular momentum of the magnetospheric plasma.
- Investigate magnetotail dynamics and substorm activity.
- Study reconnection signatures in the magnetopause and tail.
- Characterize the plasma input to the magnetosphere from the rings.
- Characterize the role of ring–magnetosphere interaction in ring particle dynamics and erosion.
- Study dust–plasma interactions and evaluate the role of the magnetosphere in species transport between Saturn's atmosphere and rings.

RADIO SCIENCE INSTRUMENT

Occultation experiments use one-way measurements. The radio signal is generated by an ultrastable oscillator aboard the spacecraft. The spacecraft radio frequency subsystems condition and amplify the signal for transmission to Earth. After passing through the "feature" of interest, the signal is detected via a parabolic DSN antenna, amplified and saved for later analysis.

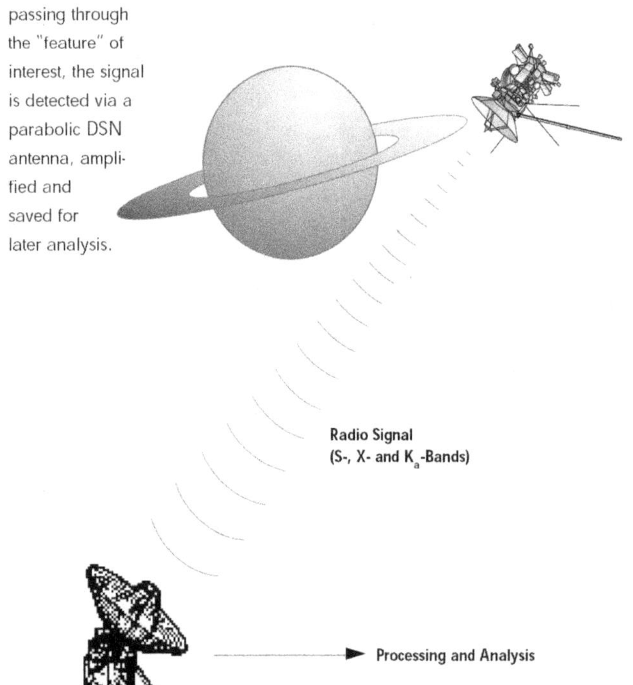

- Study the interaction of the magnetosphere with Titan's upper atmosphere and ionosphere.
- Evaluate particle precipitation as a source of Titan's ionosphere.
- Characterize plasma input to the magnetosphere from icy satellites.
- Study the effects of satellite interaction on magnetospheric particle dynamics inside and around the satellite flux tube.

The CAPS will measure the flux (the number of particles arriving at the CAPS per square centimeter per second per unit of energy and solid angle) of ions in Saturn's magnetosphere. The CAPS consists of three sensors: an electron spectrometer, an ion beam spectrometer and an ion-mass spectrometer. A motor-driven actuator rotates the sensor package to provide 208-degree scanning in azimuth about the –y-symmetry axis of the Cassini Orbiter.

FIELDS, PARTICLES AND WAVES

Six instruments designed to study magnetic fields, particles and waves are mounted in a variety of locations on the spacecraft. The fields and particles pallet holds the Cassini Plasma Spectrometer (CAPS), an Ion and Neutral Mass Spectrometer (INMS) and two additional components of the Magnetospheric Imaging Instrument (MIMI). The Cosmic Dust Analyzer (CDA) and the Radio and Plasma Wave Science (RPWS) antennas and the magnetic search coils, also part of the RPWS, are attached to various locations on the upper equipment module. The RPWS also includes a Langmuir probe. The Dual Technique Magnetometer sensors are located on an extendible boom.

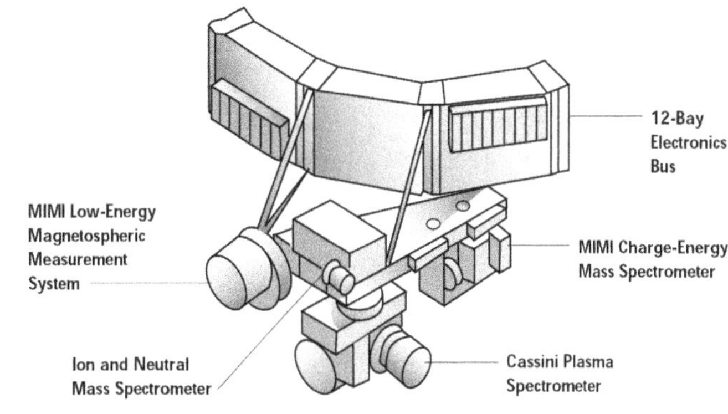

The electron spectrometer makes measurements of the energy of incoming electrons; its energy range is 0.7–30,000 electron volts. The sensor's FOV is 5 × 160 degrees. The ion-beam spectrometer determines the energy-to-charge ratio of an ion; its energy range is 1 electron volt to 50 kilo–electron volts; its FOV is 1.5 × 160 degrees. The ion-mass spectrometer provides data on both the energy-to-charge and mass-to-charge ratios. Its mass range is 1–60 atomic mass units (amu); its energy range is 1 electron volt to 50 kilo–electron volts; its FOV is 12 × 160 degrees.

Ion and Neutral Mass Spectrometer. The Ion and Neutral Mass Spectrometer (INMS) will determine the composition and structure of positive ion and neutral species in the upper atmosphere of Titan and magnetosphere of Saturn, and will measure the positive ion and neutral environments of Saturn's icy satellites and rings. The science objectives are as follows:

- Measure the in situ composition and density variations, with altitude, of low-energy positive ions and neutrals in Titan's upper atmosphere.
- Study Titan's atmospheric chemistry.
- Investigate the interaction of Titan's upper atmosphere with the magnetosphere and the solar wind.
- Measure the in situ composition of low-energy positive ions and neutrals in the environments of the icy satellites, rings and inner magnetosphere of Saturn, wherever densities are above the measurement threshold and ion energies are below about 100 electron volts.

The INMS will determine the chemical, elemental and isotopic composition of the gaseous and volatile components of the neutral particles and the low-energy ions in Titan's atmosphere and ionosphere, Saturn's magnetosphere and the ring environment. It will also determine gas velocity.

The principal components of the INMS include an open source (for both ions and neutral particles), a closed source (for neutral particles only) and a quadrupole mass analyzer and detector.

The open source analyzes ions as they enter, but the neutral gases must first be ionized within the instrument by the impact of electrons from an electron gun. The closed source measures neutrals along the direction of incoming atomic and molecular species. In both ion sources, the neutrals are ionized by electron impact.

Ions emerging from the ion sources are directed into the quadrupole mass analyzer; as the ions exit the analyzer, they go into an ion detector and are counted.

The FOV of the open source for neutral species is 16 degrees; the closed source has a hemispherical FOV. The mass range of the INMS is 1–8 amu and 12–99 amu.

Cosmic Dust Analyzer. The Cosmic Dust Analyzer (CDA) will provide direct observations of small ice or dust particles in the Saturn system in order to investigate their physical, chemical and dynamic properties and study their interactions with the rings, icy satellites and magnetosphere of Saturn. The CDA science objectives are as follows:

- Extend studies of interplanetary dust (sizes and orbits) to the orbit of Saturn.
- Define dust and meteoroid distribution (sizes, orbits, composition) near the rings.
- Map the size distribution of ring material in and near the known rings.
- Analyze the chemical compositions of ring particles.

The Cassini Plasma Spectrometer (CAPS).

The Ion and Neutral Mass Spectrometer (INMS).

The Cosmic Dust Analyzer (CDA).

- Study the erosional and electromagnetic processes responsible for E-ring structure.
- Search for ring particles beyond the known E-ring.
- Study the effect of Titan on the Saturn dust complex.
- Study the chemical composition of icy satellites from studies of ejecta particles.
- Determine the role of icy satellites as a source of ring particles.
- Determine the role of dust as a charged-particle source or sink in the magnetosphere.

The CDA measures the flux, velocity, charge, mass and composition of dust and ice particles in the mass range from 10^{-16} to 10^{-6} gram. It has two types of sensors, high-rate detectors and a dust analyzer. The two high-rate detectors, intended primarily for measurements in Saturn's rings, count impacts up to 10,000 per second, giving the integral flux of dust particles above the mass threshold of each detector. The threshold varies with impact velocity, but since ring particle orbits are nearly circular, their velocity at a given point does not vary much.

The dust analyzer determines the electric charge carried by dust particles, the flight direction and impact speed and the mass and chemical composition, at rates up to one particle per second, and for speeds of 1–100 kilometers per second. Some dust particles strike a separate chemical analyzer in the dust analyzer and also produce impact ionization. A grid accelerates positive ions; ions reaching the grid signal the start time for the time-of-flight mass spectrometer.

Other dust particles pass through into the spectrometer and eventually reach the ion collector and electron multiplier. Their time of flight is an inverse function of the ion mass. The distribution of ion masses (atomic weights) gives the chemical composition of the dust particle. The mass resolution of the ion spectrum is approximately 50.

An articulation mechanism allows the entire CDA instrument to be rotated or repositioned, relative to the Cassini Orbiter body.

Dual Technique Magnetometer. The Dual Technique Magnetometer (MAG) will determine planetary magnetic fields and study dynamic interactions in the planetary environment. The MAG science objectives include the following:

- Determine the internal magnetic field of Saturn.
- Develop a three-dimensional model of Saturn's magnetosphere.
- Determine the magnetic state of Titan and its atmosphere.
- Derive an empirical model of Titan's electromagnetic environment.
- Investigate the interactions of Titan with the magnetosphere, magnetosheath and solar wind.
- Survey ring and dust interactions with the electromagnetic environment.
- Study the interactions of the icy satellites with Saturn's magnetosphere.
- Investigate the structure of the magnetotail and the dynamic processes therein.

The MAG consists of direct-sensing instruments that detect and measure the strength of magnetic fields in the vicinity of the spacecraft. The MAG comprises both a flux gate magnetometer and a vector/scalar helium magnetometer. The flux gate magnetometer is used to make vector field measurements. The vector/scalar helium magnetometer is used to make vector (magnitude and direction) and scalar (magnitude only) measurements of magnetic fields.

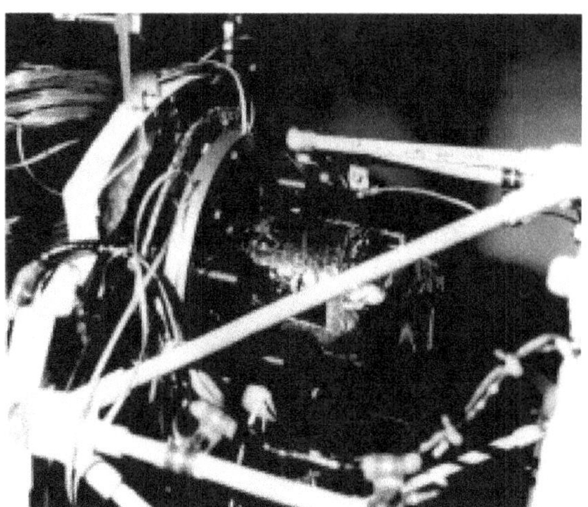

The Dual Technique Magnetometer (MAG) comprises the flux gate magnetometer (above) and the vector/scalar helium magnetometer (left).

Since magnetometers are sensitive to electric currents and ferrous components, they are generally placed on an extended boom, as far from the spacecraft as possible. On Cassini, the flux gate magnetometer is located

CASSINI ORBITER INSTRUMENTS

	Measurements	Techniques	Partner Nations
Optical Remote Sensing			
Composite Infrared Spectrometer – CIRS	High-resolution spectra, 7–1000 µm	Spectroscopy using 3 interferometric spectrometers	US, France, Germany, Italy, UK
Imaging Science Subsystem – ISS	Photometric images through filters, 0.2–1.1 µm	Imaging with CCD detectors; 1 wide-angle camera (61.2 mr FOV); 1 narrow-angle camera (6.1 mr FOV)	US, France, Germany, UK
Ultraviolet Imaging Spectrograph – UVIS	Spectral images, 0.055–0.190 µm; occultation photometry, 2 ms; H and D spectroscopy, 0.0002-µm resolution	Imaging spectroscopy, 2 spectrometers; hydrogen–deuterium absorption cell	US, France, Germany
Visible and Infrared Mapping Spectrometer – VIMS	Spectral images, 0.35–1.05 µm (0.073-µm resolution); 0.85–5.1 µm (0.166-µm resolution); occultation photometry	Imaging spectroscopy, 2 spectrometers	US, France, Germany, Italy
Radio Remote Sensing			
Cassini Radar – RADAR	K_u-band RADAR images (13.8 GHz); radiometry resolution less than 5 K	Synthetic aperture radar; radiometry with a microwave receiver	US, France, Italy, UK
Radio Science Instrument – RSS	K_a-, S- and X-bands; frequency, phase, timing and amplitude	X- and Ka-band transmissions to Cassini Orbiter; K_a-, S- and X-band transmissions to Earth	US, Italy
Particle Remote Sensing & In Situ Measurement			
Magnetospheric Imaging Instrument – MIMI	Image energetic neutrals and ions at less than 10 keV to 8 MeV per nucleon; composition, 10–265 keV/e; charge state; directional flux; mass spec: 20 keV to 130 MeV ions; 15 keV to greater than 11 MeV electrons, directional flux	Particle detection and imaging; ion-neutral camera (time-of-flight, total energy detector); charge energy mass spectrometer; solid-state detectors with magnetic focusing telescope and aperture-controlled ~45° FOV	US, France, Germany
In Situ Measurement			
Cassini Plasma Spectrometer – CAPS	Particle energy/charge, 0.7–30,000 eV/e; 1–50,000 eV/e	Particle detection and spectroscopy; electron spectrometer; ion-mass spectrometer; ion-beam spectrometer	US, Finland, France, Hungary, Norway, UK
Cosmic Dust Analyzer – CDA	Directional flux and mass of dust particles in the range 10^{-16}–10^{-6} g	Impact-induced currents	Germany, Czech Republic, France, The Netherlands, Norway, UK, US
Dual Technique Magnetometer – MAG	B DC to 4 Hz up to 256 nT; scalar field DC to 20 Hz up to 44,000 nT	Magnetic field measurement; flux gate magnetometer; vector–scalar magnetometer	UK, Germany, US
Ion and Neutral Mass Spectrometer – INMS	Fluxes of +ions and neutrals in mass range 1–66 amu	Mass spectrometry	US, Germany
Radio and Plasma Wave Science – RPWS	E, 10 Hz–2MHz; B, 1 Hz–20 kHz; plasma density	Radio frequency receivers; 3 electric dipole antennas; 3 magnetic search coils; Langmuir probe current	US, Austria, France, The Netherlands, Sweden, UK

midway out on the magnetometer boom, and the vector/scalar helium magnetometer is located at the end of the boom. The boom itself, composed of thin, nonmetallic rods, will be folded during launch and deployed long after the spacecraft separates from the launch vehicle and shortly before the Earth flyby. The magnetometer electronics are in a spacecraft bay.

The use of two separate magnetometers at different locations aids in distinguishing the ambient magnetic field from that produced by the spacecraft. The helium magnetometer has full-scale flux ranges of 32–256 nanoteslas in vector mode and 256–16,000 nanoteslas in scalar mode. The flux gate magnetometer's flux range is 40–44,000 nanoteslas.

Magnetospheric Imaging Instrument. The Magnetospheric Imaging Instrument (MIMI) is designed to measure the composition, charge state and energy distribution of energetic ions and electrons; detect fast neutral species; and conduct remote imaging of Saturn's magnetosphere. This information will be used to study the overall configuration and dynamics of the magnetosphere and its interactions with the solar wind, Saturn's atmosphere, Titan, rings and the icy satellites. The science objectives are as follows:

- Determine the global configuration and dynamics of hot plasma in the magnetosphere of Saturn.
- Monitor and model magnetospheric, substorm-like activity and correlate this activity with Saturn kilometric radiation observations.
- Study magnetosphere/ionosphere coupling through remote sensing of aurora and measurements of energetic ions and electrons.
- Investigate plasma energization and circulation processes in the magnetotail of Saturn.
- Determine through imaging and composition studies the magnetosphere–satellite interactions at Saturn, and understand the formation of clouds of neutral hydrogen, nitrogen and water products.
- Measure electron losses due to interactions with whistler waves.
- Study the global structure and temporal variability of Titan's atmosphere.
- Monitor the loss rate and composition of particles lost from Titan's atmosphere due to ionization and pickup.
- Study Titan's interaction with the magnetosphere of Saturn and the solar wind.
- Determine the importance of Titan's exosphere as a source for the atomic hydrogen torus in Saturn's outer magnetosphere.
- Investigate the absorption of energetic ions and electrons by Saturn's rings and icy satellites.
- Analyze Dione's exosphere.

The MIMI will provide images of the plasma surrounding Saturn and determine ion charge and composition. The MIMI has three sensors that perform various measurements: the low-energy magnetospheric measurement system (LEMMS), the charge-energy-mass spectrometer (CHEMS) and the ion and neutral camera (INCA).

The LEMMS will measure low- and high-energy proton, ion and electron angular distributions (the number of particles coming from each direction). The LEMMS sensor is mounted on a

The Magnetospheric Imaging Instrument (MIMI) low-energy magnetospheric measurement system, LEMMS (above), and the MIMI charge-energy-mass spectrometer, CHEMS (left).

scan platform that is capable of turning 180 degrees. The sensor provides directional and energy information on electrons from 15 kilo–electron volts to 10 mega–electron volts, protons at 15–130 mega–electron volts and other ions from 20 kilo–electron volts to 10.5 mega–electron volts per nucleon. The LEMMS head is double-ended, with oppositely directed FOVs of 15 and 45 degrees.

The CHEMS uses an electrostatic analyzer, a time-of-flight mass spectrometer and microchannel plate detectors to measure the charge and composition of ions at 10–265 kilo–electron volts per electron. Its mass/charge

The MIMI ion and neutral camera, INCA (right), and the Radio and Plasma Wave Science (RPWS) instrument (below).

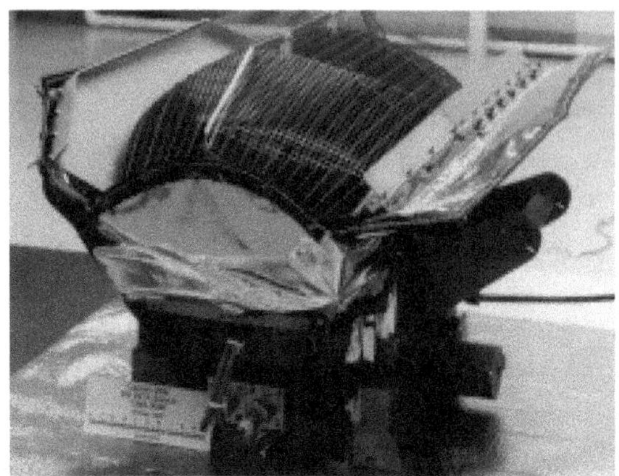

range is 1–60 amu/e (elements hydrogen–iron); its molecular ion mass range is 2–120 amu.

The third MIMI sensor, the INCA, makes two different types of measurements. It will obtain three-dimensional distributions, velocities and the rough composition of magnetospheric and interplanetary ions with energies from 10 kilo–electron volts to about eight mega–electron volts per nucleon for regions with low energetic ion fluxes. The instrument will also obtain remote images of the global distribution of the energetic neutral emission of hot plasmas in the Saturn magnetosphere, measuring the composition and velocities of those energetic neutrals. The FOV is 120 × 90 degrees.

The MIMI sensors share common electronics and provide complementary measurements of energetic plasma distribution, composition and energy spectrum, and the interaction of that plasma with the extended atmosphere and satellites of Saturn.

Radio and Plasma Wave Science. The Radio and Plasma Wave Science (RPWS) instrument will measure the electrical and magnetic fields in the plasma of the interplanetary medium and Saturn's magnetosphere, as well

as electron density and temperature. Science objectives of the RPWS instrument include the following:

- Study the configuration of Saturn's magnetic field and its relationship to Saturn kilometric radiation (SKR).
- Monitor and map the sources of SKR.
- Study daily variations in Saturn's ionosphere and search for outflowing plasma in the magnetic cusp region.
- Study radio signals from lightning in Saturn's atmosphere.
- Investigate Saturn electric discharges (SED).
- Determine the current systems in Saturn's magnetosphere and study the composition, sources and sinks of magnetospheric plasma.
- Investigate the dynamics of the magnetosphere with the solar wind, satellites and rings.
- Study the rings as a source of magnetospheric plasma.
- Look for plasma waves associated with ring spoke phenomena.
- Determine the dust and meteroid distributions throughout the Saturn system and interplanetary space.
- Study waves and turbulence generated by the interaction of charged dust grains with the magnetospheric plasma.
- Investigate the interactions of the icy satellites and the ring systems.
- Measure electron density and temperature in the vicinity of Titan.
- Study the ionization of Titan's upper atmosphere and ionosphere and the interactions of the atmosphere and exosphere with the surrounding plasma.
- Investigate the production, transport, and loss of plasma from Titan's upper atmosphere and ionosphere.
- Search for radio signals from lightning in Titan's atmosphere, a possible source for atmospheric chemistry.
- Study the interaction of Titan with the solar wind and magnetospheric plasma.
- Study Titan's vast hydrogen torus as a source of magnetospheric plasma.
- Study Titan's induced magnetosphere.

The RPWS instrument will be used to investigate electric and magnetic waves in space plasma at Saturn. The solar wind is a plasma; plasma may be "contained" within magnetic fields (that is, the magnetospheres) of bodies such as Saturn and Titan. The RPWS instrument will measure the electric and magnetic fields in the interplanetary medium and planetary magnetospheres and will directly measure the electron density and temperature of the plasma in the vicinity of the spacecraft.

The major components of the RPWS instrument are an electric field sensor, a magnetic search coil assembly and the Langmuir probe. The electric field sensor is made up of three deployable antenna elements mounted on the upper equipment module of the Cassini Orbiter. Each element is a collapsible beryllium copper tube that is rolled up during launch and subsequently unrolled to its 10-meter length by a motor drive.

The magnetic search coil assembly includes three orthogonal coils about 25 millimeters in diameter and 260 millimeters long. The magnetic search coils are mounted on a small platform attached to a support for the HGA. The Langmuir probe, which measures electron density and temperature, is a metallic sphere 50 millimeters in diameter. The probe is attached to the same platform by a one-meter deployable boom.

Signals from the sensors go to a number of receivers, which provide low and high time and frequency resolution measurements. The instrument ranges are one hertz to 16 megahertz for electric fields; one hertz to 12.6 kilohertz for magnetic fields; electron densities of 5–10,000 electrons per cubic centimeter; and electron temperatures equivalent to 0.1–4 electron volts.

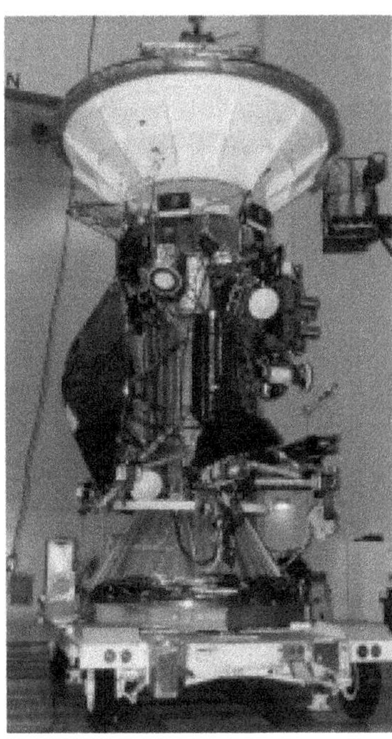

The Cassini–Huygens spacecraft, with thermal blankets on, in its flight-ready state.

Probe Instruments Overview

The Huygens Probe, provided by the European Space Agency (ESA), will fly to the Saturn system aboard the Cassini–Huygens spacecraft. Carrying six science instruments, the Probe will study the atmosphere and surface of Saturn's largest satellite, Titan, descending through Titan's atmosphere and landing — either on solid land or in a liquid lake or ocean.

Many countries have participated in the development of the Huygens Probe's six instruments. The instruments are the Huygens Atmospheric Structure Instrument (HASI), the Aerosol Collector and Pyrolyser (ACP), the Gas Chromatograph and Mass Spectrometer (GCMS), the Descent Imager and Spectral Radiometer (DISR), the Doppler Wind Experiment (DWE) and the Surface Science Package (SSP). These instruments comprise 39 sensors; the total mass of the science payload is 48 kilograms.

Once the Probe has separated from the Orbiter, it is wholly autonomous. The instruments are turned on in a preprogrammed sequence after the Probe's heat shield is released. Probe instrument operation is controlled by timers, acceleration sensors, altimeters and a Sun sensor. All instruments use time-based operation from the beginning of the descent down to 20 kilometers. Three of the instruments, the DISR, the HASI and the SSP, use measured altitude below 20 kilometers. Up to three hours of Probe data will be relayed to the Orbiter for later transmission to Earth.

Probe Instrument Descriptions

Huygens Atmospheric Structure Instrument. The Huygens Atmospheric Structure Instrument (HASI) investigates the physical and electrical properties of Titan's atmosphere, including temperature, pressure and atmospheric density as a function of altitude and wind gusts — and in the event of a landing on a liquid surface — wave motion. Comprising a variety of sensors, the HASI will also measure the ion and electron conductivities of the atmosphere and search for electromagnetic wave activity.

Atmospheric pressure is measured by the deflection of a diaphragm. A tube inlet is mounted on a stub extending outside the Probe. Thermometers mounted on the stub measure atmospheric temperature.

HUYGENS PROBE INSTRUMENTS

	Measurements	Techniques	Partner Nations
Huygens Atmospheric Structure Instrument – HASI	Temp: 50–300 K; Pres: 0–2000 mbar; Grav: 1 μg–20 mg; AC E-field: 0–10 kHz, 80 dB at 2 $\mu Vm^{-1} Hz^{-0.5}$; DC E-field: 50 dB at 40 mV/m; electrical conductivity: 10^{-15} Ω/m to ∞; relative permittivity: 1–∞; acoustic: 0–5 kHz, 90 dB at 5 mPa	Direct measurements using "laboratory" methods	Italy, Austria, Finland, Germany, France, The Netherlands, Norway, Spain, US, UK
Gas Chromatograph and Mass Spectrometer – GCMS	Mass range: 2–146 amu; Dynamic range: $>10^8$; Sensitivity: 10^{-12} mixing ratio; Mass resolution: 10^{-6} at 60 amu	Chromatography and mass spectrometry; 3 parallel chromatographic columns; quadrupole mass filter; 5 electron impact sources	US, Austria, France
Aerosol Collector and Pyrolyser – ACP	2 samples: 150–45 km, 30–15 km altitude	3-step pyrolysis; 20°C, 250°C, 650°C	France, Austria, Belgium, US
Descent Imager and Spectral Radiometer – DISR	Upward and downward spectra: 480–960 nm, 0.87–1.7 μm; resolution 2.4–6.3 nm; downward and side-looking images: 0.66–1 μm; solar aureole photometry: 550 nm, 939 nm; surface spectral reflectance	Spectrometry, imaging, photometry and surface illumination by lamp	US, Germany, France
Doppler Wind Experiment – DWE	(Allan Variance)1/2: 10^{-11} (in 1 s), 5×10^{-12} (in 10 s), 10^{-12} (in 100 s), corresponding to wind velocities of 2 m/s to 200 m/s, Probe spin	Doppler shift of Huygens Probe telemetry signal, signal attenuation	Germany, France, Italy, US
Surface Science Package – SSP	Gravity: 0–100 g; Tilt: ±60°; Temp: 65–100 K; thermal conductivity: 0–400 mW $m^{-1}K^{-1}$; speed of sound: 150–2000 m/s; liquid density: 400–700 kg m^{-3}; refractive index: 1.25–1.45	Impact acceleration; acoustic sounding, liquid relative permittivity, density and index of refraction	UK, Italy, The Netherlands, US

To determine the density of the atmosphere, an accelerometer measures acceleration along the spin axis. Three additional accelerometers measure acceleration along all three axes of the Probe over a range of ±20 g. A microphone is also part of the HASI; it senses acoustic noise from sources such as thunder, rain or wind gusts.

A permittivity and wave analyzer, consisting of an array of six electrodes, is mounted on two deployable booms. Measurements of the magnitude and phase of the received signal give the permittivity and electronic conductivity of the atmosphere and the surface. Electromagnetic signals such as those of lightning can also be detected. The instrument also processes the signal from the Probe's radar altimeter to obtain information on surface topography, roughness and electrical properties.

Aerosol Collector and Pyrolyser. The Aerosol Collector and Pyrolyser (ACP) captures aerosol particles from Titan's atmosphere using a deployable sampling device extended in the air flow below the nose of the Probe. The samples are heated in ovens to vaporize the volatiles and decompose the complex organic materials. The products are then passed to the GCMS for analysis.

The ACP will obtain samples at two altitude ranges. The first sample, at

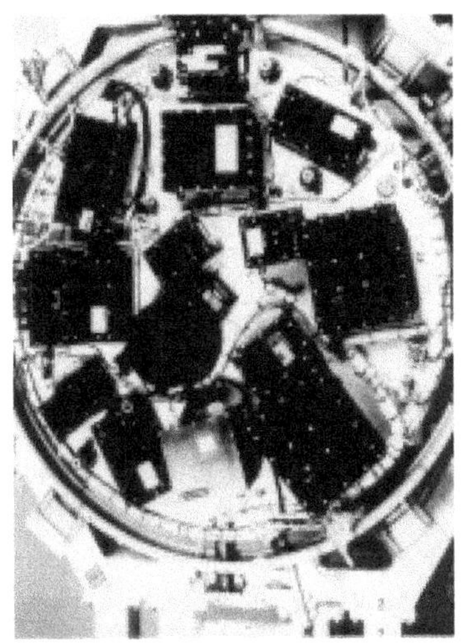

altitudes down to 40 kilometers above the surface, will be obtained primarily by direct impact of the atmosphere on a cold filter target. The second sample will be obtained at about 20 kilometers by pumping the atmosphere through the filter.

After each collection, the filter will be transferred to an oven and heated to three successively higher temperatures, up to about 650 degrees Celsius, to vaporize and pyrolyze the collected material. The product at each temperature will be swept up by nitrogen carrier gas and transferred to the GCMS for analysis.

Gas Chromatograph and Mass Spectrometer. The Gas Chromatograph and Mass Spectrometer (GCMS) provides a quantitative analysis of the composition of Titan's atmosphere. Atmospheric samples are transferred into the instrument by dynamic pressure as the Probe descends through the atmosphere. The Mass Spectrometer constructs a spectrum of the molecular masses of the gas driven into the instrument. Just prior to landing, the inlet port of the GCMS is heated to vaporize material on contact with the surface. Following a safe landing, the GCMS can determine Titan's surface composition.

The GCMS uses an inlet port to collect samples of the atmosphere and has an outlet port at a low pressure point. The instrument contains three chromatographic columns. One column has an absorber chosen to separate carbon monoxide, nitrogen and other gases. Another column has an absorber that will separate nitriles and other highly polar compounds. The third is to separate hydrocarbons up to C_8. The mass range is 2–146 amu.

The Mass Spectrometer serves as the detector for the Gas Chromatograph, for unseparated atmospheric samples and for samples provided by the ACP. Portions of the GCMS are identical in design to the Orbiter's INMS.

Descent Imager and Spectral Radiometer. The Descent Imager and Spectral Radiometer (DISR) uses several instrument fields of view and 13 sensors, operating at wavelengths of 350–1700 nanometers, to obtain a variety of imaging and spectral observations. The thermal balance of the atmosphere and surface can inferred by measuring the upward and downward flux of radiation.

Solar aureole sensors will measure the light intensity around the Sun resulting from scattering by aerosols, permitting calculations of the size and number density of suspended particles. Infrared and visible imagers will observe the surface during the latter stages of the descent. Using the Probe's rotation, the imagers will build a mosaic of pictures of the Titan landscape. A side-looking visible imager will view the horizon and take pictures of the clouds, if any exist. For spectral measurements of the surface, a lamp will be turned on shortly before landing to provide enough light for measuring surface composition.

The DISR will obtain data to help determine the concentrations of atmospheric gases such as methane and argon. DISR images will also determine if the local surface is solid or liquid. If the surface is solid, DISR will reveal topographic details. If the surface is liquid, and waves exist, DISR will photograph them.

DISR sensors include three framing imagers, looking downward and horizontally; a spectrometer dispersing light from two sets of optics looking downward and upward; and four solar aureole radiometers. The spectral range of the imagers is 660–1000 nanometers; the spectrometer's range is 480–960 nanometers; the aureole radiometers operate at 475–525 and 910–960 nanometers, with two different polarizations.

Separate downward- and upward-looking optics are linked by fiberoptic bundles to an infrared grating spectrometer. The infrared detectors have a spectral range of 870–1700 nanometers. There are also two violet photometers, looking downward and upward with a bandwidth of 350–470 micrometers.

To provide reference and timing for the other measurements, the DISR uses a Sun sensor to measure the solar azimuth and zenith angle relative to the rotating Probe.

The interior of the Huygens Probe, showing its science instruments.

Doppler Wind Experiment. The Doppler Wind Experiment (DWE) uses two ultrastable oscillators (USOs), one on the Probe and one on the Orbiter, to give Huygens' relay link a stable carrier frequency. Orbiter measurements of the shift in Probe frequency (Doppler shift) will provide information on the Probe's motion from which a height profile of the zonal wind (the component of wind along the line of sight) and its turbulence can be derived.

The output frequency of each USO is set by a rubidium oscillator. The signal (which is like a very high precision clock) received from the Probe is compared with that of the Orbiter USO, and the difference in frequency is recorded and stored for transmission to Earth. This information is used to determine the Doppler velocity between the Probe and the Orbiter. Modulation of this signal will provide additional data on the Probe spin rate, spin phase and parachute swing. Winds will be measured to a precision of one meter per second.

Surface Science Package. The Surface Science Package (SSP) contains a number of sensors to determine the physical properties and composition of Titan's surface. Some of the SSP sensors will also perform atmospheric measurements during descent.

During descent, measurements of the speed of sound will give information on atmospheric composition and temperature. An accelerometer records the deceleration profile at impact, indicating the hardness of the surface. Tilt sensors (liquid-filled tubes with electrodes) will measure any pendulum motion of the Probe during descent, indicate the Probe orientation after landing and measure any wave motion.

If the surface is liquid, an opening at the bottom of the Probe body, with a vent extending upward along the Probe axis, will admit liquid, which will fill the space between a pair of electrodes. The capacitance between the electrodes gives the dielectric constant of the liquid; the resistance gives the electrical conductivity. A float with electrical position sensors determines the liquid's density.

A sensor to measure the refractive index of the liquid has light-emitting diode (LED) light sources, a prism with a curved surface and a linear photodiode detector array. The position of the light/dark transition on the detector array indicates the refractive in-

PROBING TITAN'S DEPTHS

Like the Cassini Orbiter, Huygens represents an international collaboration. The partners involved in the Probe effort are Austria, Belgium, the European Space Agency, Finland, France, Germany, Italy, The Netherlands, Norway, Spain, the United Kingdom and the United States. The Huygens science instruments comprise a total of 39 sensors. The total mass of the Probe science payload is 48 kilograms. Once the Probe separates from the Orbiter, it is wholly autonomous. The instruments will turn on in a preprogrammed sequence after the Probe cover is released. Probe operation is controlled by timers, acceleration sensors, altimeters and a Sun sensor. As much as three hours of Probe data will be relayed to the Orbiter and later transmitted to Earth.

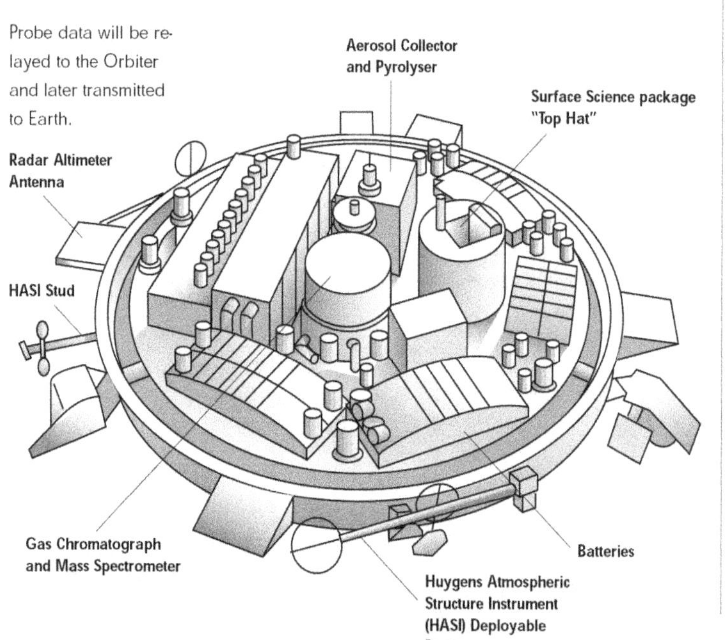

dex. A group of platinum resistance wires, two of which can be heated, will measure temperature and thermal conductivity of the surface and lower atmosphere and the heat capacity of the surface material. The acoustic sounder, which will then work as a sonar, will conduct an acoustic sounding of liquid depth — if the Probe lands in liquid.

The Huygens Probe is released on a trajectory to enter Titan's atmosphere.

Instrument Operations

Cassini Orbiter. The power available to the Cassini Orbiter is not sufficient to operate all the instruments and engineering subsystems simultaneously. Operations are therefore divided into a number of operational modes. For example, during much of the orbital tour of Saturn, 16 hours in remote-sensing mode will often alternate with eight hours in fields and particles and waves and downlink mode.

Other science modes will be used during satellite flybys, occultations, cruise to Saturn and so on. Instruments that are not gathering data will generally not turn off during the orbital tour, but will be in a low-power "sleep" state. This strategy is designed to reduce on–off thermal cycling, keep high voltages on (to avoid having to turn up voltage slowly each time it is required) and to preserve the onboard computer memories (to avoid having to reload them each time).

The operational modes differ in characteristics other than power. In remote sensing, for example, the Orbiter is oriented to point remote-sensing instruments toward their objects of interest. This means that the HGA generally cannot be pointed toward Earth, so telemetry is stored in the solid-state recorders for later transmission.

In fields, particles, waves and downlink mode, the HGA is pointed toward Earth — permitting transmission of stored and real-time telemetry — and the Orbiter is rolled about the antenna axis to provide scanning about another axis in addition to the articulation axes of some instruments.

The bit rate available on the command and data subsystem databus is not high enough to permit all instruments to output telemetry simultaneously at their maximum rates. The Orbiter is switched among a number of different telemetry modes in which the available bit rate is allocated differently among the instruments.

Some of the instruments will adjust their operating state or parameters depending on the activities of the spacecraft or other instruments and the kind of environment the Orbiter is encountering. When these other conditions are predictable from the command sequence, commands for the instruments will be set accordingly.

When conditions are not predictable, or if it is simpler to handle the adjustment on board, the command and data subsystem relays information to the instruments that need it. Specifically, information on spacecraft attitude and its rate of change, warnings of thruster firings, measurements of the magnetic field vector and notices of operation of the sounder and Langmuir probe in the RPWS instrument are broadcast for use by the CAPS, CDA and MIMI.

Huygens Probe. There is no radio transmission link to the Huygens Probe after it separates from the Orbiter; the Probe is wholly autonomous. The instruments are turned on in a preprogrammed sequence after the Probe cover is released. The Probe goes through five successive power configurations in which the available power is allocated differently among the various instruments. Operation is controlled by timers, acceleration sensors, altimeters and a Sun sensor. The Probe's command and data management subsystem broadcasts altitude and spin-rate data to the instruments. Data collection involves three successive steps in which the available data rate is allocated differently among the instruments.

Summary

The Cassini–Huygens mission will carry 18 scientific instruments to the Saturn system. After the spacecraft is inserted into Saturn orbit, it will separate into a Saturn Orbiter and an atmospheric Probe, called Huygens, which will descend to the surface of Titan. The Orbiter will orbit the planet for four years, with close flybys of Enceladus, Dione, Rhea, Hyperion and Iapetus, and multiple close flybys of Titan.

The Orbiter is three-axis stabilized. Its 12 science instruments are body-mounted; the spacecraft must be turned to point them toward objects of interest. Optical instruments provide imagery and spectrometry at wavelengths from 55 nanometers to one millimeter. A radar instrument supplies synthetic aperture imaging, altimetry and microwave radiometry. S-band, X-band and K_a-band link measurements between the Orbiter and Earth will provide information about intervening material and gravity fields. Fields and particles instruments will measure magnetic and electric fields, plasma properties and the flux and properties of dust and ice particles. Cassini's maximum downlink rate from Saturn to Earth is 166 kilobits per second.

The Probe is spin-stabilized for the coast to Titan. A heat shield decelerates it and protects it from heat during entry into Titan's atmosphere. Parachutes then slow its descent to the surface and provide a stable platform for taking images and making other measurements. A set of small vanes on the bottom skirt of the Probe forces it to rotate, providing 360-degree viewing of the landscape.

The Probe carries six instruments, including sensors to determine atmospheric physical properties and chemical composition. Radiometric and optical sensors will provide data on temperatures and thermal balance and obtain images of Titan's atmosphere and surface. Doppler measurements over the radio link from Probe to Orbiter will provide wind profiles. Surface sensors are carried to measure impact acceleration, thermal properties of the surface material and, if the surface is liquid, its density, refractive index, electrical properties and acoustic velocity. The Probe returns its data via an S-band link to the Orbiter.

CHAPTER 10

An Ensemble Effort

Mission operations are the ensemble of actions to plan and execute the launch and subsequent activities of a spacecraft, including data return. And since that data flow is divided into uplink and downlink, operations processes also fall into these two categories. Uplink encompasses planning mission activities; developing and radiating instrument and spacecraft commands, and execution of those commands. Downlink encompasses data collection; transmission to the ground, and data processing and analysis for system performance evaluation and science studies.

General Mission Operations

Mission operations involve complex interactions among people, computers, electronics, and even heavy machinery (at NASA's Deep Space Network antenna sites). Planning for mission "ops" begins with the conception of the mission itself, and involves assumptions about the frequency of communication with the spacecraft and data transfer rates and volumes. These, in turn, drive the design of the spacecraft hardware and software and of the ground system that will be manipulating the spacecraft during its interplanetary cruise and mission operations phases.

Sequence Planning

After basic design decisions have been made, their effects on day-to-day and long-term operations are considered, and a system is developed to enable spacecraft engineering, health and safety requirements and science data acquisition requirements to be met. The period of mission operations, starting shortly after launch, is divided into a series of intervals. These intervals may be constrained by engineering requirements (e.g., limitations on available spacecraft memory), mission events (e.g., a planetary flyby), flight rules (e.g., the spacecraft must have communications with the ground every 240 hours) or other constraints.

A "sequence" is prepared for each planning interval. The sequence is a series of computer commands that tells the spacecraft and its engineering and science subsystems what to do and when. These commands cover everything from the mundane, such as turning a heater on, to the sublime, such as aiming the camera and taking a sequence of approach images to make into a movie.

Uplink and Downlink

Most of the human interaction occurs at the beginning of sequence planning, when the push-pull of spacecraft engineering requirements and science measurement requirements and desires must be resolved. (Note that the push-pull of competing science requests must have been resolved first.) After what may be long hours of negotiations, the interests of all the concerned parties are met (more or less) and the actual work of sequence preparation can begin. This involves first generating human-readable commands (for proofreading purposes) and then converting them into the computer code words that the space-

"Aces," as mission controllers are known, provide round-the-clock monitoring of deep space exploration missions such as Cassini–Huygens.

craft can understand. The computer codes are then packaged into radio-transmissible "tones," instructions are given to the antenna on where to point, and the sequence is radiated to the spacecraft. The transmission is received and decoded by the spacecraft, which sends an acknowledgment and executes the commands. This portion of mission operations is referred to as uplink.

Downlink is what follows. In the course of executing the commands, both engineering subsystems and science instruments generate data that are often stored on the spacecraft for a while before they are coded and packaged for transmission to the ground as telemetry. On Earth, the data are decoded and forwarded to the engineering and science teams concerned with those particular activities on the spacecraft. The data are analyzed and engineering reports and science publications are generated for dissemination to their respective communities.

Signals Through Space

None of this happens instantaneously nor easily. Electromagnetic radiation — whether it is visible light with wavelengths in the range of 400 to 700 nanometers (400–700 billionths of a meter) or radio waves with wavelengths from 1 millimeter to 30 kilometers or longer (frequencies of 300 gigahertz to 10 kilohertz or lower) — take time to travel through space. The velocity of electromagnetic radiation in a vacuum is 299,792 kilometers per second. One-way light-times in excess of a few seconds (the equivalent distance to the Moon is about 1.3 seconds) make "joysticking" the spacecraft impractical. While provision is made for realtime commands, especially planned ones and those needed in response to emergencies, preplanned sequences are almost always used for commanding the spacecraft.

The amount of electromagnetic radiation (emitted by a point source) passing through a unit area decreases as the square of the distance. Thus, a spacecraft twice as far as Earth is from the Sun receives only one-fourth of the light and heat that it would receive at Earth's distance.

What is true of light and heat is true of radio transmissions as well, and it affects how telecommunications are accomplished. The radio transmitters and receivers on the ground and on the spacecraft are equipped with paraboloidal antennas that direct transmissions to the receiver and concentrate received radio energy, maximizing signal to the preamplifier–receiver–amplifier chain.

Large antennas are a necessity. While a commercial radio station may generate as much as 50,000 watts in its signal to your car radio a few tens or hundreds of kilometers away, a spacecraft typically has only a 20-watt transmitter that must reliably send signals over hundreds of millions or billions of kilometers across the solar system.

Spacecraft communications are typically accomplished using radio frequencies in S-band (2–4 gigahertz) or X-band (7–12 gigahertz). Some radio science measurements made with a spacecraft — for example, tests of Einstein's general theory of relativity — are made in K_a-band (32–34 gigahertz). (Radio astronomical observations from the ground and from spacecraft are made across a wide region of the spectrum.)

A Worldwide Effort

Cassini operations involve numerous people at many sites around the world. Careful coordination of the efforts of the science, engineering, planning, navigation and sequencing teams will permit the Orbiter and Probe to return a cache of scientific data on Saturn and its environs that will increase our understanding manyfold and allow scientists many years of additional study after the end of the mission.

Cassini–Huygens Operations

The purpose of Cassini–Huygens mission operations is to launch an instrumented spacecraft to orbit Saturn and deliver a Probe to Titan in order to accomplish primary scientific objectives. The program is a joint undertaking among NASA, the European Space Agency (ESA) and the Italian space agency, Agenzia Spaziale Italiana (ASI).

Mission Description

Cassini will be launched aboard a Titan IVB/Centaur launch vehicle with Solid Rocket Motor Upgrades and injected into a 6.7-year trajectory to

PLANETARY MISSION OPERATIONS THE JPL WAY

Function	Primary Activities
Mission Planning	Generate a mission plan and mission phase plans. Generate mission and flight rules and constraints.
Integrated Sequence Development	Generate detailed timelines of activities. Generate integrated files of valid commands. Generate sequence and realtime command loads for radiation to the spacecraft.
Mission Control	Configure and control the ground system. Monitor in real time the spacecraft's and payload's activities, health and safety.
Navigation	Predict and reconstruct the spacecraft's trajectory. Design maneuvers to correct the trajectory to achieve mission objectives.
Spacecraft Analysis and Sequence Input Development	Plan, design and integrate engineering activities. Maintain the health and safety of the spacecraft through non-realtime analyses and anomaly identification and resolution.
Payload Analysis and Sequence Input Development	Generate a science operations plan. Design and integrate instrument observations. Maintain the health and safety of the payload through non-realtime analyses and anomaly identification and resolution.
Science Processing	Generate and archive processed data in support of payload analysis and for the purpose of scientific research.
Database Management	Provide throughout the mission realtime accessible engineering and science telemetry and ancillary data for non-realtime spacecraft and instruments analyses.
Data Acquisition	Acquire the downlink signal. Generate Doppler and ranging tracking data. Extract and digitize the original information from the modulated subcarrier of the radio signal. Eliminate bit errors.
Digital Processing	Perform telemetry decommutation.* Display and convert telemetry channels data.
Data Transport	Ensure that communications are in place and functioning so that data can be routed to the project database and scientists.
Computer and Communication Administration	Implement and administer computer and communication systems.
Training	Train personnel on the processes and on the use of the ground data system.
System Engineering	Design a mission operations system that matches the requirements, takes constraints into account and has acceptable performance.

**Decommutation is the process by which a single telemetry signal is separated into its component signals.*

Saturn, gathering energy from two flybys of Venus, one flyby of Earth and one of Jupiter. Upon arrival at Saturn, the spacecraft will be placed into orbit around Saturn and a four-year tour of the Saturn system will begin.

The Huygens Probe will be delivered on the first (nominal) or second (backup) flyby of Titan. Multiple close Titan flybys will be used during the tour for gravity assists and studies of the satellite. The tour also includes icy satellite flybys for satellite studies and orbits at a variety of inclinations and orientations with respect to the Sun–Saturn line for ring, atmospheric, magnetospheric and plasma science studies.

Spacecraft and Payload

The Cassini spacecraft is a three-axis stabilized spacecraft. The Huygens Probe and the Orbiter instruments are affixed to the body of the spacecraft. There are 18 science instrument subsystems, which are divided into four groups: optical remote sensing; microwave remote sensing; fields, particles and waves; and Probe instruments.

Orbiter instruments that serve multiple investigations are called facility instruments. Facility instruments are provided by the Jet Propulsion Laboratory (JPL), the NASA Goddard Space Flight Center or by JPL and ASI. The facility instruments, except for the Ion and Neutral Mass Spectrometer (INMS), are operated by a JPL team called the distributed operations interface element.

Instruments that serve individual investigations are provided and operated by a Principal Investigator (the INMS is operated like a Principal Investigator instrument). The Huygens Titan Probe is operated from the Huygens Probe Operations Centre in Darmstadt, Germany.

Capabilities and Constraints

The Cassini–Huygens mission scientific objectives involve complex mission, spacecraft and payload demands. In addition, mission operations must fit within certain inherent constraints. However, operational capabilities built into the spacecraft can help to keep operations simple.

Operational Capabilities

Robustness. Robustness is provided by redundant engineering subsystem assemblies, healthy margins of Orbiter consumables and fault-protection software that provides protection for the spacecraft and the mission in the event of a fault.

Automation. Automation is provided by the onboard inertial vector propagator, which generates turn profiles, bringing the spacecraft from its last commanded attitude and rate to a possibly time-varying target attitude and rate. The target attitude is obtained by applying an offset to a base attitude, which is defined by primary and secondary pointing constraints as follows:

- Primary constraint — Align a spacecraft fixed vector (primary body vector) with a specified inertial vector (primary inertial vector).
- Secondary constraint — Align a spacecraft fixed vector (secondary body vector) with a specified inertial vector (secondary inertial vector).

Flexibility. Flexibility is provided by the capability of storing telemetry data on a solid-state recorder for later playback and by trigger-command capabilities. Most instruments operate from commands stored in their memories; a trigger command is issued from the central computer to initiate a preestablished series of instrument-internal commands.

Inherent Constraints

Cost. The major operational constraint of the Cassini–Huygens mission is that mission operations and data analysis activities must be conducted within fixed budgets (within both the U.S. and the European space communities).

Thermal. At Sun range closer than 2.7 astronomical units (AU), thermal constraints make it necessary to use the Cassini's high-gain antenna to shade the spacecraft from the Sun — except for infrequent turns away from the Sun (off-Sun turns), such as those used for trajectory correction maneuvers. The allowable duration of the off-Sun turns roughly scales with the square of the Sun range. For example, at 0.61 AU, the spacecraft could withstand a transient off-Sun duration of 30 minutes.

Telecommunications. While the high-gain antenna is used to shade the spacecraft, telecommunications mostly will be restricted to one of two low-gain antennas[1] and will proceed at very low bit rates (generally less than 40 bits per second).

Power. The electrical power from Cassini's radioisotope thermoelectric generators (676 watts at the beginning of the tour and 641 watts at the end of the tour) is not sufficient to operate all instruments and engineering subsystems simultaneously.

Data Rates. The bit rate available on the command and data subsystem data bus — which is limited to a total of 430 kilobits per second (kbps) — is not high enough to permit all instruments to output telemetry simultaneously at their maximum rates. Instruments and engineering subsystems also must share the number of bits stored on solid-state recorders and transmitted to the ground. During the tour, expected data rates are on the order of 14–166 kbps.

Pointing. The major pointing constraint arises from the fact that the instruments are fixed to the body of the spacecraft. Other pointing constraints prevent exposing sensitive spacecraft and payload components to undesirable thermal input. The pointing accuracy is two milliradians when the spacecraft is not rotating and pointing toward a fixed inertial direction. For target-relative pointing, the accuracy is limited by navigational uncertainties.

Navigation. During the tour, the spacecraft's orbit as a whole is controlled by Saturn's gravity. For satellite flybys, preflyby tracking is needed to support the maneuver that places the spacecraft in the final flyby trajectory. Postencounter maneuvers are required only for Titan as a result of its gravitational effect on the trajectory. A compromise must be reached to balance the desire to acquire science data during Titan's flyby with the desire of performing the post-encounter maneuver soon after closest approach in order to minimize the amount of propellant used for that maneuver.

Propellant. This is the principal consumable of the mission. At launch, the spacecraft will carry 2919 kilograms of bipropellant for the main engine and 132 kilograms of hydrazine for the thrusters. Once the bipropellant is gone, no significant maneuver can be planned (unless significant leftover hydrazine is used). Once the hydrazine is gone, the spacecraft will begin to lose attitude control. For this reason, propellant budgets are subject to particular scrutiny.

The Probe. After separation from the Orbiter, the Probe will be powered by five batteries. After 22 days of coasting, there will be enough power to operate the Probe for three hours from entry, including two and a half hours of descent and 30 minutes on the surface. The data rate over the Probe–Orbiter link will be 16 kbps. During cruise, Probe checkouts must be performed to verify the capability of executing the Probe mission. The checkouts simulate as closely as possible the sequence of activities to be performed when the Probe approaches Titan.

BRINGING IMAGES FROM SPACE TO EARTH

NASA brings us unforgettable images of space. In addition to their enormous scientific value, these images dazzle and capture our imagination. Bringing images from space to Earth – in effect, doing long-distance photography – is a complicated process that depends on our ability to communicate with spacecraft millions of kilometers from Earth. This communication is the responsibility of NASA's Deep Space Network (DSN), a global system of powerful antennas.

Taking the picture is the job of the spacecraft's imaging system: digital camera, computer and radio. Reflected light from the target – for example, Saturn's rings or the large satellite Titan – passes through the lens and one or more color filters before reaching a charge-coupled device (CCD). The surface of the CCD is made up of thousands of light-sensitive picture elements, or "pixels." Each pixel assigns a number value for the light it senses; the values are different for each color filter. The spacecraft's computer converts the data into digital code – bits – which are transmitted to Earth. The bit stream is received by huge antenna receivers at one of the three DSN sites: Goldstone, California; Canberra, Australia; and Madrid, Spain. The data are then sent to JPL in Pasadena, California, where the bits are reformatted, calibrated and processed to ensure a true representation of the target, then recorded on high-quality black-and-white or color film for archiving and distribution.

Of course, the DSN must handle not only image data but all the other kinds of data Cassini transmits to Earth as well. For flexibility, Cassini will use a variety of antenna configurations – 34-meter, 70-meter or an array of 34- and 70-meter dishes.

Operational Choices

The following operational choices were made to reduce costs and simplify mission operations.

Limited Science

To reduce costs, there is no plan to acquire science data during inner and outer cruise phases. The only exception is a gravitational wave experiment which, following the Jupiter flyby, will attempt to detect gravitational waves emitted by supermassive dynamical objects such as quasars, active galactic nuclei or binary black holes.

Science data will not be collected during the planetary flybys[2] except for calibrations of the Cassini Plasma Spectrometer, the Dual Technique Magnetometer and the Magnetospheric Imaging Instrument at Earth, and calibration of the Radio and Plasma Wave Science instrument at Jupiter. Other science activities during the inner and outer cruise phases will be limited to deployments, maintenance, characterizations and checkout. Science observations will begin two years before Saturn orbit insertion.

Limited Operational Selections

To simplify operations, the number of allowable operational selections in several areas of mission operations has been reduced to a limited set of configurations.

Telemetry Modes. A set of telemetry modes has been defined to provide rates for recording data on the solid-state recorder and for downlink in the context of changing telecommunication capabilities. Each telemetry mode represents a unique configuration of data sources, rates and destinations for telemetry data gathered and distributed by the command data subsystem. There are five types of telemetry data: engineering, science housekeeping, scientific, playback from the solid-state recorder, and Probe. The telemetry modes are grouped into nine functional categories. The three primary modes for operations during the orbital tour are realtime engineering, science and engineering record and realtime engineering and science playback.

Operational Modes. The concept of operational modes was invented to reduce operational complexity arising from the facts that there is insufficient power to operate all the instruments simultaneously, and that all the instruments cannot be optimally pointed in a simultaneous fashion because they are fixed to the body of the spacecraft.

The operational modes concept prescribes that instruments will operate in a series of standard, well-characterized configurations. Each operational mode is characterized by the state of each instrument (for example, "on," "off," "sleep," etc.); minimum and maximum power and peak data rate allocations for each instrument, and states of certain engineering subsystems: the radio frequency subsystem, the solid-state recorder, the attitude and articulation control subsystem and the propulsion module subsystem.

WHAT GOES UP...MUST COME DOWN

This diagram shows the uplink and downlink data flow process. (Housekeeping data provide information about science instruments.)

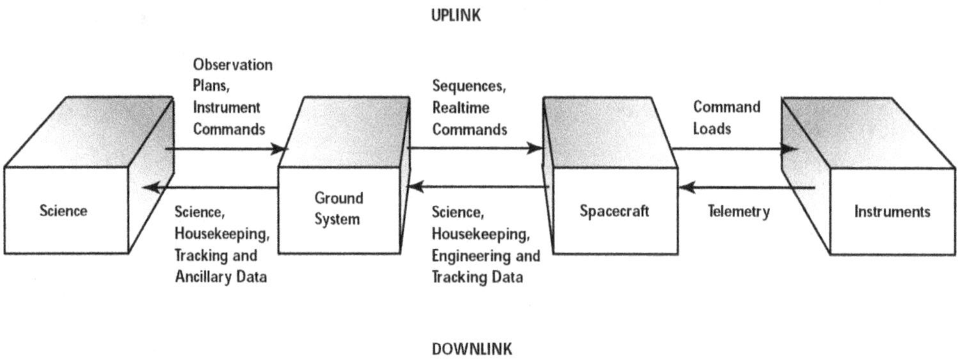

UPLINK

Science

Observation Plans, Instrument Commands

Science, Housekeeping, Tracking and Ancillary Data

Ground System

Sequences, Realtime Commands

Science, Housekeeping, Engineering and Tracking Data

Spacecraft

Command Loads

Telemetry

Instruments

DOWNLINK

During much of Cassini's orbital tour, 15 hours in an optical remote sensing mode will alternate with nine hours in a fields, particles, waves and downlink mode. Other modes will be used for radar and radio science observations. Instruments that are not collecting data are generally left in a low-power sleep state, so as to reduce on–off thermal cycling, keep high voltages on and avoid reloading instrument memories each time they need to operate.

Basic Mission. This concept prescribes that 98 percent of the tour will be conducted with a small number of reusable multi-instrument sequence modules and templates. The remaining two percent (seven days a year) may consist of unique sequences. These sequence constructs are defined as follows:

- A module is a reusable science sequence of commands integrating system-level and trigger commands, pointing functions and telemetry mode selections.[3]
- A template is a sequential series of fixed-duration modules, fixed sequences, gaps of fixed durations or other templates whose relative timing is set.
- A fixed sequence is of fixed duration and is designed and validated once for multiple uses, does not use operational modes and has a fixed list of variable inputs.
- A unique sequence has a specific purpose, is used once and does not use modules.

Data Return. Deep Space Network (DSN) coverage of Cassini during the tour will consist of one pass per day, with occasional radio science passes. To reduce operational complexity, the coverage is divided into high-activity and low-activity days, with four gigabits or one gigabit of data returned per day, respectively. A quarter of each orbit, or seven days, whichever is less, will be considered high-activity.

To simplify operations while maintaining some flexibility for data gathering, three configurations for data return have been selected. They differ essentially by the DSN antennas used (34-meter antennas, 70-meter antennas or 34-meter and 70-meter antennas arrayed) and the durations of the passes.

Tour Maneuvers. Propulsive maneuvers during the orbital tour will generally occur three days before and two days after each Titan flyby, near apoapsis (the farthest point in the spacecraft's orbit).

Workforce Management

Distributed Operations. Distributed operations places observing decisions, including generation of instrument-internal subsequences, in the hands of the science teams. The implementation of distributed operations for the Cassini mission is achieved through computers, computer-resident software and communication lines provided by JPL to the remote sites, as well as science participation in the uplink (mission planning, sequence development) and downlink (Principal Investigator instrument health monitoring) processes.

Virtual Teams. Cassini uses virtual teams for mission planning and sequence development. These teams bring together people for the development of a given product. Their membership varies, depending on the particular subphase or sequence to be developed. Each virtual team member maintains membership in his/her mission team of origin and continues to work for that team while providing expertise to and generating products for the virtual team.

Cassini–Huygens will launch on October 6, 1997, from Cape Canaveral in Florida, aboard a Titan IVB/Centaur launch vehicle.

Uplink Processes

Cassini mission operations uses five uplink processes: science planning, mission planning, sequence development, realtime command development and radiation.

Science Planning

The purpose of the science planning process is to plan conflict-free science activities. During the inner and outer cruise phases, the science planning process is particularly simplified because the acquisition of science data is limited to deployments, maintenance, calibrations, checkout and the gravitational wave experiment.

In preparation for the science cruise phase and the tour, Cassini scientists will prepare a science operations plan. The science office and the four discipline working groups — atmospheres, magnetospheric and plasma science, rings and satellite surfaces — will play a major role in coordinating this effort.

Mission Planning

The mission planning process is the responsibility of the mission planning virtual team. This process focuses on particular subphases of the mission and consists of two subprocesses: mission plan and mission phase plans updates, and phase update package generation.

Mission Plan and Mission Phase Plans Updates. The mission plan is the principal reference for a high-level description of the Cassini–Huygens mission. It documents spacecraft design and trajectory, mission phases, high-level activities and operational strategies for collection of scientific and engineering data. The plan serves as a common starting point and a guide for the implementation of mission operations.

Mission phase plans describe specific mission phases. Examples include the "Cassini Spacecraft System Maintenance, Calibration and Deployment Handbook," which identifies those activities listed in the document's title as well as other "one-time only"[4] or science activities of the inner and outer cruise phases, and the Cassini Inner Cruise Activity Plan, which describes all spacecraft activities from two days after launch until the end of the inner cruise phase. These plans are updated for a given mission subphase to provide levels of detail adequate for implementation. Changes in spacecraft performance or in ground capabilities may make it necessary for the mission planning virtual team to perform trade studies and mission analyses.

Phase Update Package Generation. This activity consists of translating the updated sections of the plans into a set of inputs, called a Phase Update Package, for the sequence development team(s). A major portion of this task consists of developing timelines for each sequence of the mission subphase. Activities contained in those timelines include spacecraft events, geometric events, data rates, solid-state recorder strategies and DSN allocations, with a resolution time between one hour and one day.

Sequence Development

The sequence development process is the responsibility of the sequence virtual team. This process focuses on particular sequences and consists of three subprocesses: activity planning, subsequence generation and sequence integration and validation.

Activity Planning. The sequence timeline is refined to one-minute resolution. Inclusion of the latest DSN allocations, revisions arising from the verification of resource allocations and other revisions such as timing changes are included in the timeline. The timeline is verified for compliance with activity level mission and flight rules and constraints.

Subsequence Generation. Science, engineering, system-level and ground-event subsequences are implemented to the command level. The subsequences are generated by the appropriate teams. For instance, the team known as the flight system operations element generates the subsequences for activities, the distributed operations interface element generates the subsequences for the facility instruments, the Orbiter Principal Investigator teams generate the subsequences for their instruments and the Huygens Probe Operations Centre commands the Probe (prior to its separation from the spacecraft).

Sequence Integration and Validation. The set of subsequences is merged into an integrated sequence. The sequence virtual team leader verifies that the syntax is correct and that mission and flight rules and constraints are not violated. There may be a need to validate the sequence through simulation, which can be done in the Cassini high-speed simulator or in the Cassini integration test laboratory.

The high-speed simulator is a high-fidelity simulator of the command data subsystem and attitude and articulation control subsystem's computers. The integration test laboratory contains hardware identical to hardware flown on the spacecraft, in particular the command and data subsystem, solid-state recorders, attitude and articulation control subsystem's computers, sensors and actuators.

Realtime Commands

A realtime command is one that is generated and transmitted subsequent to the nominal sequence development process. It can be executed immediately after receipt by the spacecraft or its execution may be delayed. There are several types of realtime commands. Some are planned and pregenerated, and are used for repetitive transmissions. They are not part of the sequence, but the sequence contains a window to send the realtime command if needed. Some realtime commands are planned but not generated, and are generally used for updating parameters. The need to change the parameter value is known, but the value to which it should be changed is not. Realtime commands that are unplanned, with varying degrees of time criticality, are emergency commands.

System-level realtime commands are handled by the sequence virtual team. There is a plan to have instrument-internal realtime commands handled automatically after the requester has generated, verified and delivered the realtime command file to the project's central database.

Radiation

After translation into appropriate format by a JPL team called the realtime operations element, sequence and command files are sent or "radiated" to the spacecraft by a DSN antenna.

Downlink Processes

Cassini–Huygens mission operations use six downlink processes: realtime data processing, data storage and distribution, realtime monitoring, non-realtime data analysis, science data analysis and science data archiving.

The flyby of Earth in Cassini–Huygens' primary trajectory occurs on August 16, 1999, 57 days after the second Venus flyby. This artist's rendition shows the spacecraft over South America.

Realtime Data Processing

A deep space station antenna acquires the signal transmitted by the spacecraft, passes it through a low-noise amplifier, which boosts the energy level of the signal, and distributes it to the signal processing center.

In the signal processing center, the signal goes through a receiver, where it is converted into an electrical signal and the subcarrier is separated from the carrier. (The carrier, which is the main component of the radio signal, is modulated with information-carrying variations; the subcarrier is a modulation applied to the carrier.) The receiver distributes the subcarrier and the tracking data to separate assemblies. In the telemetry string, the telemetry is separated from the subcarrier and digitized, the digitized encoded data are corrected to eliminate bit errors (loss of information by a change in value of a bit by some chance effect) and they are blocked into frame-size records.

The full records are then routed via the ground communication facility interface to the telemetry input subsystem, where science, instrument housekeeping and engineering data are separated. The realtime operations element operates the telemetry input subsystem.

Data Storage and Distribution

The science packets and instrument housekeeping and engineering channelized data are stored on the telemetry delivery system. Selected data are broadcast for realtime monitoring and displayed by the data monitor and display program. For non-realtime analysis, data are obtained through queries of the telemetry delivery system.

Realtime Monitoring

The mission controllers, engineering teams and science teams monitor telemetry and look for anomalies in real time. They process the channelized data, looking for evidence of anomalous behavior, displaying the data in various ways and performing special processing.

Non-Realtime Data Analysis

The flight system operations element retrieves engineering data used to determine the health, safety and performance of the spacecraft, and processes the tracking data to determine and predict the spacecraft's trajectory. The team generates ancillary information that is delivered to the project central database.

The distributed operations interface element for the facility instruments, the Principal Investigator teams for their instruments and the operations team for the Probe process the data, converting telemetry and raw tracking data into products usable for verification that planned activities have been performed as expected and of instrument health and safety evaluation and science data analysis.

Science Data Analysis

The Cassini science team members and the interdisciplinary scientists (who use scientific information from two or more instruments provided by science teams) are responsible for this process. Science data analysis is the process by which scientific studies using the data products are conducted to evaluate the information content of the measurements, characterize physical phenomena, identify causes and effects, appraise existing theories and develop new ones — and produce highly processed data and publications to increase and disseminate scientific knowledge.

Science Data Archiving

This is the process by which data are recorded and stored in approved national data archives for future references and scientific studies.

FOOTNOTES

[1] At approximately 14 months after launch, the Earth–spacecraft–Sun angle will be small enough to allow use of the high-gain antenna for telecommunications. That 25-day period will be used for a checkout of all the Orbiter instruments. This will be the first real opportunity after launch for the science teams to verify that their instruments are working properly.

[2] Science at Jupiter is currently not planned, but is not precluded.

[3] The inertial vector propagation and trigger command capabilities are key to enabling the implementation of modules.

[4] An example of a one-time only activity is the launch plus 14 months Orbiter instrument checkout.

It has been said that in making scientific and technological advances, each generation of scientists and engineers stands on the shoulders of the previous generation. This is true of all types of exploration, and it is true with the Cassini–Huygens mission.

On the Shoulders of Giants

The story begins in late 1609, when Galileo Galilei built a telescope and had the inspiration to aim it at the sky. Galileo's discoveries about the Moon, Venus, Jupiter and the Milky Way were a quantum leap forward in our understanding of the universe, but he was puzzled by Saturn. With his telescope's poor optics, Galileo could only see that Saturn looked like a "triple" planet. To add to the mystery, the planet lost its "companions" a few years later!

Fast forward more than 40 years: Christiaan Huygens turned his telescope toward Saturn and discovered a real companion to the planet – like the four discovered around Jupiter by Galileo – a satellite now called Titan. Huygens also recognized for the first time the great ring surrounding Saturn. He realized that the plane of the ring is tilted with respect to the plane of the planet's orbit, so that about every 14 years, Earth crosses through the plane and the ring can disappear temporarily – explaining Galileo's puzzling observation. Shortly thereafter, Jean-Dominique Cassini discovered four additional satellites and spied a "gap" in the ring, making it two rings separated by what is now called the Cassini division.

Over the next century, more satellites were discovered. And two centuries later, James Clerk Maxwell, well-known for his explanation of electromagnetism, showed that the rings were not solid disks but were instead composed of myriad particles, each in its own orbit. Astrophysical techniques, in development since the 1860s, enabled in 1932 the discovery of methane and ammonia in Saturn's atmosphere – and thus a great advancement in understanding the nature of Saturn and the other "gas giants" – Jupiter, Uranus and Neptune. This was followed by Gerard Kuiper's surprising discovery in 1943–44 of a dense atmosphere, unknown among other planetary satellites, on Titan.

The birth of space exploration in 1957, and the National Aeronautics and Space Administration in 1958, led to dreams of human and robotic exploration of the planets. The wanderlust of the Renaissance was reborn as both scientists

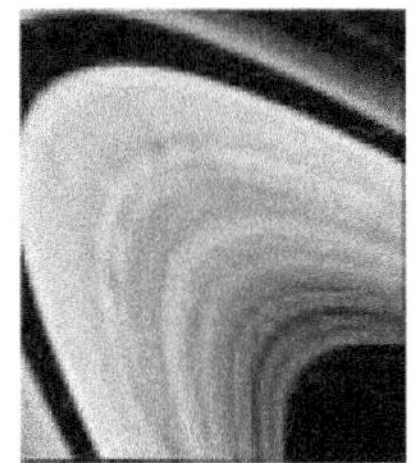

and the public became interested in visiting distant worlds. Ground-based planetary astronomy experienced a rebirth, as new techniques and the need for more information about the planets to be explored pushed the science along.

The evolution of planetary exploration began with the first steps to the Moon, was followed by the exploration of Venus and Mars and continued in plans for exploring the outer planets (that is, Jupiter and beyond). Fanciful "Grand Tour" missions to the outer planets were proposed. Pioneers 10 and 11 explored Jupiter and its environment in 1973 and 1974, respectively. The success of Pioneer 10 permitted the retargetting, after Jupiter, of Pioneer 11 to arrive at Saturn in 1979, just as Voyager 1 was arriving at Jupiter.

The Voyagers, 1 and 2, carried out a thorough reconnaissance of Jupiter and then Saturn. Following the success at Saturn, Voyager 2 was able to complete the Grand Tour, missing only tiny Pluto in its travels.

The contributions of these missions, complemented by the continuing developments of science and technology, have brought us to the Cassini–Huygens mission. We can only guess, and not well, at what the Cassini–Huygens spacecraft will find. Surely, many questions raised by previous observations will be answered. But the answers, as always in science, will only generate more questions. And, it is safe to predict there will be some unimaginable surprises.

Technology – the Bridge from the Ethereal to the Material

As technology has improved through history, our understanding of natural phenomena has also improved. And, this technology has affected our daily lives. Improved telescopes provided more knowledge about Saturn and Titan, generating more questions – which drove the development of technology in optics. This has not only helped us do remote-sensing science better, it has allowed eyeglass wearers to see better, via improved lens materials and coatings.

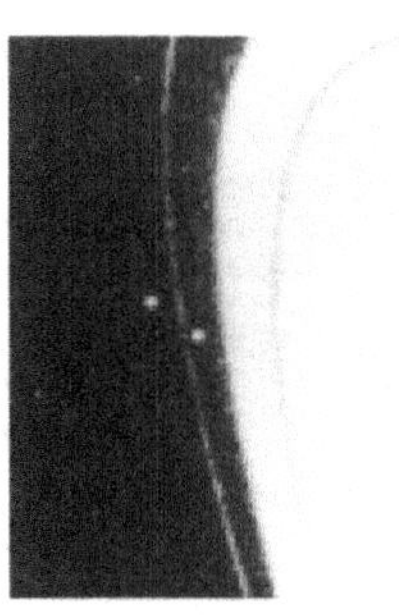

The development of auxiliary instruments changed astronomy from a descriptive science into a "measuring" science, uncovering new mysteries for us to puzzle over, like Titan's atmosphere. Improvements in detectors, from eyeballs through photography to electronic imagers, have made possible the permanent, unbiased recording of images. Digitized images can be transmitted, without loss of detail, across the solar system – and soon, directly into your home, with the image quality on your television matching that in a movie theater.

Advancements in detectors and image-analysis software have been applied in diverse areas: helping patients with limited night vision "see" in the dark; enabling doctors to analyze magnetic resonance, positron emission and X-ray images; and protecting airport personnel from exposure to radiation from security devices.

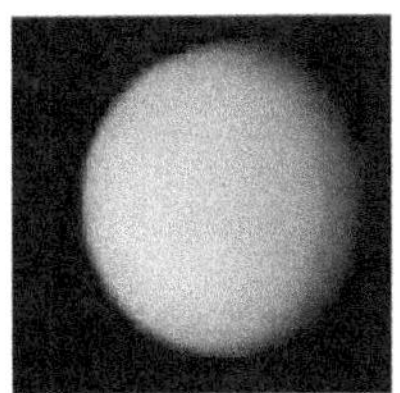

Developments tied to the Cassini–Huygens mission have had an effect on everyday life. The solid-state recorder, using silicon chips to store data, has found application on other spacecraft and in whole industries, from aerospace to entertainment. Powerful computer chips and a radio transponder from the mission are being used in other spacecraft, providing better performance and lowering overall cost. Solid-state power switches developed for the mission can be used in electrical and electronic products for both industry and consumers.

A new resource-trading exchange permits subsystems on the Cassini–Huygens spacecraft to help balance conflicting needs. It delegates any decisions to those who understand the problem best, so better decisions result. Already, this system has been used by California's South Coast Air Quality Management District to help regulate air pollution; the state of Illinois is adapting it to manage volatile organic wastes.

The Cassini Management Information System tracks a myriad of receivables and deliverables by establishing "contracts" among those involved. Critical steps and objects are easier to identify – and problems can be identified earlier, so they can be addressed more easily and with less expense.

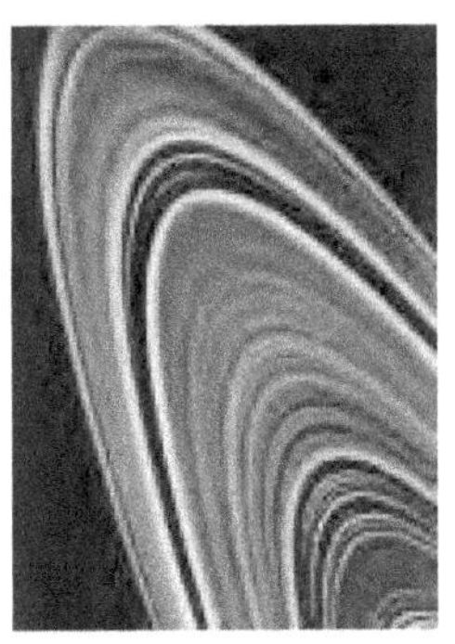

The Cassini–Huygens mission involves hundreds of scientists, engineers and technicians from 17 countries. Learning to cooperate and work together toward a common goal is good practice, for technicians as well as governments. International cooperation distributes the expense among the partner nations and gives them all the opportunity to participate in the adventure and the science. More people with different backgrounds than ever before will share in the direct and derived benefits from this mission.

Space exploration inspires us to look beyond our everyday existence and to the greater universe. We come into this world as explorers, engineers and scientists, learning initially by taste and feel and going on, first crawling, then walking, then running – as the preceding generations try to keep up! – to see and experience the world around us. Exploration is instinctive in us; the Cassini–Huygens mission is another way for us to extend our senses physically to new places and over a broader range of phenomena.

Beautiful Saturn, with its rings, its moons and its many other wonders, beckons to us to explore and uncover its secrets, to partake in the thrill of discovery. Cassini–Huygens is the vehicle.

APPENDIX A

Glossary of Terms

A

Accrete Gravitationally accumulate mass; the process by which planets form.

Aeronomy. The science of the physics of upper atmospheres.

Aerosols. Large molecules or particulates that remain suspended in air over long periods of time.

Albedo. The fraction of incident radiation reflected by a planet or satellite. If the value is integrated over all wavelengths, it is called the bolometric albedo. If the value is integrated over all directions, it is called the Bond albedo. The geometric albedo is the ratio of the brightness at a phase angle of zero degrees (full illumination) compared with a diffuse, perfectly reflecting disk of the same size.

Apparition. The duration of visibility of a planet during any given year.

Apoapsis. The farthest point in an orbit from the body being orbited.

Arrayed antennas. Ground antennas made to work together in order to enhance the telemetry capability.

Astronomical unit (AU). The mean distance from the Sun to Earth – 149,597,870.694 kilometers.

Aureole A luminous area surrounding the Sun or other bright light when seen through a thin atmosphere.

B

Bit error A loss of information that occurs as a result of the change in value (1 to 0, or 0 to 1) of a single bit by some chance effect.

Black hole. An object whose gravitational field is so strong that even light cannot escape from it.

Bow shock A standing shock wave that forms upstream of a planet. The bow shock forms because the planet's magnetic field creates an obstacle to the incoming solar wind flow. The bow shock heats the solar wind and slows it to subsonic speeds so that it can flow around the planet's magnetosphere.

C

Carbonaceous (C-type) material. Carbon silicate primordial material rich in simple organic compounds. C-type material is spectrally flat and exists on the surfaces of several outer planet satellites and C-type asteroids.

Carrier The main component of a radio signal generated by a transmitter.

Central Data Base. A database administered by the Cassini Program that stores relevant project-specific data such as sequence files and ancillary data.

Clathrate. A chemical compound consisting of a lattice molecule and inclusions of a smaller molecule within the crystal lattice; ice or other substance that traps molecules in its crystal structure without chemical bonding.

Very Large Array, Socorro, New Mexico

Command. A data transaction which, when issued to a recipient, causes an activity to take place within that recipient's hardware, software or both. A command is defined by a name and input parameters. Commands in a sequence are time-tagged.

Comminuted satellite A satellite (moon) reduced to tiny fragments.

Conjunction. Occurs when the direction from Earth to the spacecraft is the same as that to the Sun.

Convection Transport of energy by means of motion of the material (generally refers to charged particles moving through a magnetic field, gases moving through surrounding atmosphere, or vertical motions of liquids or partially melted solids within a satellite).

Corotating. Moving with the planet's rotation (generally refers to the charged particles in Saturn's magnetosphere that are being carried along by Saturn's magnetic field, which rotates with the planet).

Cosmic rays. Very energetic atomic nuclei, the most energetic of which come from outside of our solar system. Cosmic rays can enter planetary magnetospheres and interact with particles there.

D

DSN pass. An interval of time during which a Deep Space Network (DSN) station is used to communicate with a spacecraft.

D-type material. Primordial, low-albedo, reddish material believed to be rich in organic compounds. It exists on D-type asteroids and on the surfaces of some of Saturn's satellites.

Data block. A segment of data with fixed sizes, organized with a specific structure with a header containing a variety of codes, such as spacecraft identification or source of data.

Data packet A group of data that has been structured into standardized, discrete units for transmission by the spacecraft.

Deceleration Diminishing speed.

Deuterium Heavy hydrogen, so named because the nucleus of a deuterium atom contains a neutron in addition to the proton carried by ordinary hydrogen atoms.

Differentiation Melting and chemical fractionation of a planet or satellite into a core and mantle; the gravitational separation of different kinds of material in different layers in the interior of a planet, generally with denser layers deeper, as a result of heating.

Jean-Dominique Cassini, astronomer

Doppler data Data that determine the speed at which a spacecraft is approaching or departing Earth. The Doppler data are derived from the change of frequency of a downlink signal.

Downlink. On the path from the spacecraft to the ground.

Dynamo theory The theory that attributes planetary magnetic fields to the flow of electric currents in the interior of the planet.

E

Eclipse The passage of a satellite or spacecraft through the shadow of a larger planet or satellite; the Sun is not visible from the smaller body during an eclipse.

Emission. The sending out or giving off of light, infrared radiation, radio waves or matter by a body.

Endogenic Caused by some process originating within a planet or satellite.

Engineering subsystems The spacecraft subsystems that maintain day-to-day functions of the spacecraft and support the science instruments.

Exogenic Caused by exterior processes acting on a planet or satellite.

F

Facility instrument One of the five Cassini Orbiter science instruments (INMS, ISS, RADAR, RSS and VIMS) provided by the Cassini Program for use by a selected team of scientists (as opposed to being provided by the selected team).

G

Galilean satellites The four largest satellites of Jupiter, discovered by and named for Galileo Galilei.

Galileo Galilei, scientist

Gravitational field. The region of space surrounding a planet or satellite within which gravitational forces from that planet or satellite can be detected; generally also characterized at each point by a strength and direction.

Gravitational wave. Distortion of space and time by a massive object as predicted by Einstein's general theory of relativity.

Gravity. The force of attraction between masses in the universe.

Gravity assist. Modification of a spacecraft trajectory by passage near a planet; modification of a spacecraft orbit around a planet by passage close to a satellite of that planet.

H

Housekeeping data Data that provide information about the status of a science instrument.

Hydrate A chemical compound with water ice bound to the crystal lattice or adsorbed to the crystal surface.

I

Illumination The lighting up, or the geometry associated with the lighting up, of a surface or atmosphere.

In situ The Latin meaning is "in (its original) place." Spacecraft measurements that sample the local environment make in situ measurements in contrast to cameras, which sense the environment "remotely."

Interdisciplinary Utilizing information from two or more scientific instruments to deduce the nature of some phenomenon.

Ion. An atom or molecule electrically charged by the loss or addition of one or more electrons. In planetary atmospheres or magnetospheres, most ions are positively charged, implying that one or more electrons are lost.

Ionosphere. The electrically conducting plasma region above the atmosphere of a planet (or Titan) in which many of the atoms are ionized, where charged particles (ions and electrons) are abundant.

Isotope. One of two or more forms of an element that differ in atomic mass due to differing numbers of neutrons in the nuclei of its atoms.

APPENDIX A

J

Jovian planet A planet with general characteristics similar to those of Jupiter, also referred to as a "gas giant" planet. The Jovian planets in our solar system are Jupiter, Saturn, Uranus and Neptune.

K

Kelvin A unit of a temperature scale with its zero level at absolute zero temperature (–273.16 degrees Celsius).

Kilometer Preferred distance unit in planetary studies, equal to 1000 meters; about 62 percent of a mile.

Kuiper Belt A belt of small bodies of ice and rock left over from the formation of the solar system. These bodies reside in near-circular orbits beyond Neptune.

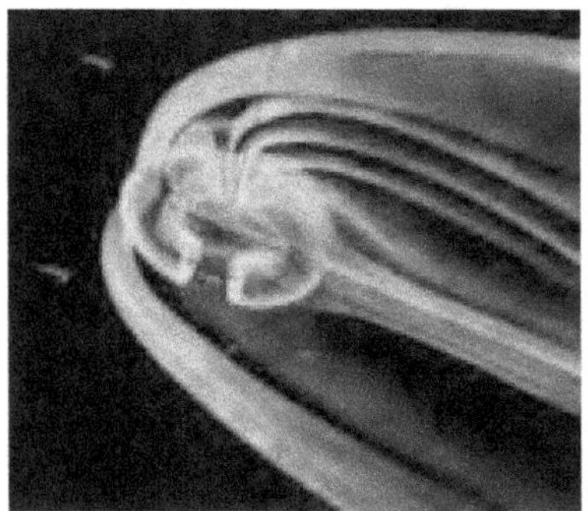

Saturn's immense magnetosphere

L

Lag deposit A residual surface deposit remaining after other components of the original surface material have been removed.

Lagrange points Equilibrium points in the orbit of a planet or satellite around its primary. Smaller bodies may reside near these points, which are about 60 degrees ahead of or behind the planet or satellite in its orbit.

M

Mach number The ratio of the speed of the solar wind to the speed of compressional waves. The Mach number of a bow shock is an indication of the bow shock's strength.

Magma Subsurface molten material that may cause volcanic activity on a moon or planet if it breaks through the surface.

Magnetic anomaly. A perturbation within an otherwise symmetric dipolar magnetic field. At Saturn, a magnetic anomaly is thought to allow charged particles to precipitate along field lines deep into the magnetosphere and account for Saturn kilometric radiation and auroras.

Magnetic field The region of space surrounding a magnetized planet or satellite in which a moving charge or magnetic pole experiences a force; generally also characterized at each point by a strength and direction.

Magnetic reconnection A process whereby magnetic field lines are "cut" and "reconnected" to different field lines, allowing the magnetic topology to change.

Magnetopause. The boundary of the magnetosphere that separates the incoming solar wind from the magnetosphere.

Magnetosheath Layer of deflected and shock-heated solar wind plasma between the bow shock and the magnetopause.

Magnetosphere The region around a planet where the planet's magnetic field and associated charged particles (plasma) dominate over the relatively weak interplanetary field carried by the solar wind.

Magnetotail The region of the magnetosphere that stretches away from the Sun (up to several hundred planetary radii) due to the drag of the solar wind flow past the planet's magnetosphere.

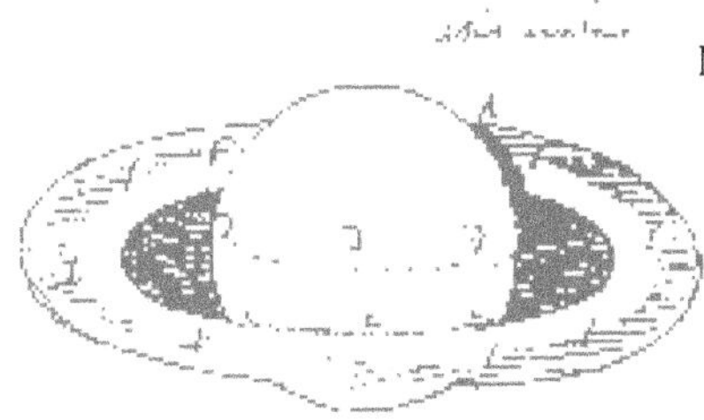

Meteorite A piece of celestial debris that hits the surface of planets or satellites; it may range in size from invisible dust to asteroid-sized chunks.

Morphology The external structure of a satellite surface in relation to form or topographic features that may lead to interpretation of the geological history of that surface.

N

Nitrile A compound containing nitrogen.

Noble gas. A nonreactive gas; specifically, helium, neon, argon, krypton, xenon or radon.

Nonresonant orbit An orbit wherein the spacecraft orbital period is not an integer multiple of the satellite's orbital period. Used for orbits that have consecutive flybys with the same satellite at different orbit locations.

Nontargeted flyby A relatively close (less than 100,000 kilometers) flyby that is not tightly controlled but occurs by serendipity during the Saturn tour.

Nucleation The process whereby raindrops condense on a preexisting solid grain of material that is suspended in the atmosphere.

O

Occultation Passage of one object behind another object as viewed by the observer or spacecraft.

Opposition Occurs when the direction from Earth to a spacecraft is the opposite of that to the Sun.

Orbit cranking Changing the orbit of a spacecraft from a gravity-assist flyby without changing its orbital period.

Orbit pumping Changing the orbital period of a spacecraft from a gravity-assist flyby.

Organic molecule A molecule containing carbon; organic molecules are not necessarily associated with life or living organisms.

Orthogonal Descriptive of two or three directions in space that are at right angles (perpendicular) to each other.

P

Packet. The data transmitted by the spacecraft are structured into standardized, discrete data packets to ease their handling and to reduce data noise.

Periapsis The point in an orbit closest to the body being orbited.

Phase angle The angle at the target between the observer (or spacecraft) and the Sun.

Photochemistry Chemical reactions caused or promoted by the action of light (usually ultraviolet light) which excites or dissociates some compounds and leads to the formation of new compounds.

Plasma A completely ionized gas, the so-called fourth state of matter (besides solid, liquid and gas), in which the temperature is too high for atoms as such to exist and which consists of free electrons and free atomic nuclei.

Plasma sheet Plasma of "hot" particles surrounding the neutral sheet in the magnetotail, where the magnetic field reverses from an orientation toward to an orientation away from the planet.

APPENDIX A

Plasma wave A wave characterized by displacement motions of ions within a plasma.

Polar caps The polar regions in the northern and southern hemispheres in which the magnetic field lines are open to the solar wind. The aurorae form at the boundaries of the polar caps.

Polymers Structures in which the same molecular configuration repeats again and again.

Primary body The celestial body (usually a planet) around which a satellite orbits.

Prograde motion The orbital motion of a planet or satellite that orbits its primary in the same direction as the rotation of the primary.

Q

Quasar Also called quasi-stellar object. An object with a dominant starlike component with spectral lines showing a large redshift.

R

Radiometric dating A method of estimating the age of an object. It entails measuring the fraction of a radioactive material that has decayed and inferring the time elapsed based on laboratory measurements of the decay rate of the radioactive substance.

Ranging data Data that determine the distance between a tracking antenna on Earth and a spacecraft, produced by impressing a ranging code on the radio signal between Earth and the spacecraft.

Reconnaissance Preliminary survey of a planetary system revealing the general characteristics of that system.

Refractivity A measure of the amount of bending experienced by a light beam traveling from one transparent material into another; related to the relative speeds of light in the two materials.

Regolith The broken or pulverized surface layer or rocky debris fragmented by meteorite impacts.

Remote sensing. Observations using instrumentation that collects data on an object from a location remote from that object.

Remote site An institution outside the Jet Propulsion Laboratory from which a Principal Investigator or a Team Leader conducts his or her investigation.

Resonance. A pattern of recurring orientations in the orbital positions or rotational states of planets or satellites, leading to repeated gravitational perturbations.

Resonance point. A radial distance within the Saturn ring system where the orbital period of the ring particles is an integer multiple of a relatively nearby satellite's orbital period.

Resonant orbit An orbit in which the spacecraft orbital period is an integer multiple of the satellite's orbital period. Used for orbits that have consecutive flybys with the same satellite at the same orbit location.

Saturn orbit insertion for Cassini

Retrograde motion. The orbital motion of a planet or satellite that orbits its primary in a direction opposite that of the rotation of the primary.

Roche limit The distance (equal to about 2.44 times the radius of the primary) within which the tidal forces exerted by the primary on an icy satellite exceed the internal gravitational forces holding the satellite together.

S

Satellite An object orbiting a planet; natural satellites are often referred to as "moons."

Saturn electrostatic discharge. Radio bursts from Saturn with a periodicity of about 10 hours, 10 minutes; thought to be due to lightning activity in Saturn's equatorial atmosphere.

Saturn kilometric radiation The major type of radio emission from Saturn, characterized by wavelengths in the kilometer range and a periodicity of about 10 hours, 39.4 minutes. The emission comes from regions in the auroral zones when these regions are near local noon.

Saturn local time. Solar time as measured for a local region on Saturn; the Sun is on the local meridian at local noon.

The Huygens Probe descends to Titan

Science payload. The assemblies used for science data collection. For Cassini, this includes the science instruments on the Orbiter and the Huygens Probe.

Sequence A computer file of valid commands used to operate the spacecraft for a predetermined period of time.

Slurry A mixture of liquid and solid material, often of the same chemical composition.

Solar wind The highly ionized plasma streaming radially outward from the Sun at supersonic speeds. It consists largely of protons and electrons in nearly equal numbers with a small amount of ionized helium and ions of heavier elements. Embedded in it is the weak interplanetary magnetic field that originates at the Sun.

Spectroscopic analysis Determination of the composition and other characteristics of a surface or gas based on the relative brightnesses at a number of different colors.

Sputtering Erosion of a satellite's surface as a result of bombardment by charged particles corotating with the magnetosphere.

Stratosphere. The layer of the atmosphere between the troposphere and the ionosphere; a stable layer in which heat is predominantly radiated (as opposed to conducted) away.

Subcarrier Modulation, applied to a carrier, which is itself modulated with information-carrying variations.

Subsequence A portion of a sequence that pertains to a single investigation or engineering subsystem.

Synchronous rotation A dynamic state caused by tidal interactions in which the satellite always presents the same face toward the primary.

Synergism The combination of results from different science studies that produces a greater amount of information than the sum of the individual results.

T

Target The object that is observed at a given time by one or more spacecraft science instruments.

APPENDIX A

Deep space burn during cruise to Saturn

Targeted flyby A flyby that passes through a specified (usually very close) aimpoint at the time of closest approach.

Tectonics. A branch of geology dealing with surface structure, especially folding and faulting.

Telecommunications The process of commanding and receiving information from a remotely located spacecraft via radio signals.

Telemetry. A stream of data bits radiated through space on electromagnetic waves or transmitted by wire or electronic pulses that represents the output of some sensor or scientific instrument.

Tenuous Very low density; difficult to detect (descriptive of E-ring particle distribution).

Tholin. Greek word meaning "muddy." Describes the orange–brown products of laboratory experiments aimed at simulating the chemistry in Titan's atmosphere.

Torus A collection of charged or neutral particles, shaped like a torus or doughnut, and generally associated with the orbital track of a satellite (like Titan).

Tracking data Data needed to track a spacecraft. These data are extracted from the properties of the radio signal received from the spacecraft. Examples of tracking data are Doppler and ranging data.

Trajectory The path of a body (i.e., a spacecraft) in space.

Transonic flow Flowing or moving near the local speed of sound.

Tropopause The boundary between the troposphere and the stratosphere.

Troposphere The atmospheric layer closest to the surface of Saturn or Titan in which convection is the dominant process for transporting heat; the layer in which "weather" takes place.

Turbulence A state of commotion or stormy agitation within an atmosphere.

U

Uplink. On the path from Earth to the spacecraft.

V

Viscosity The quality or property of a fluid or aggregate of particles that causes it to resist free flow.

Viscous relaxation. The slumping and subsequent disappearance of a feature due to gravitational forces acting on a material.

W

Wake. The track left by a body moving through a fluid.

APPENDIX B

Acronyms & Abbreviations

A

A-ring Outermost of Saturn's main rings

AACS Attitude and articulation control subsystem

ACC Accelerometer

ACP Aerosol Collector and Pyrolyser (Huygens science instrument)

AFC AACS Flight Computer

AQ60 Material used for Huygens Probe heat tiles

ASI Agenzia Spaziale Italiana, the Italian space agency

ASIC Application-specific integrated circuit

AU Astronomical unit, the mean distance from Earth to Sun = 149,597,870.694 kilometers

B

B-ring Brightest and densest of Saturn's rings

BIU Bus interface unit

C

C-ring Faint ring inward of Saturn's B-ring

C-type Asteroid material, rich in carbon

C_2H_2 Acetylene

C_2H_4 Ethylene

C_2H_6 Ethane

C_3H_8 Propane

CH_4 Methane

CAPS Cassini Plasma Spectrometer (Orbiter science instrument)

CDA Cosmic Dust Analyzer (Orbiter science instrument)

CDS Command and data subsystem

CIRS Composite Infrared Spectrometer (Orbiter science instrument)

CRAF Comet Rendezvous/Asteroid Flyby spacecraft or mission

D

D-ring Innermost Saturn ring

D-type Asteroid material, rich in hydrocarbons

DISR Descent Imager and Spectral Radiometer (Huygens science instrument)

DSN Deep Space Network (tracks operating spacecraft)

DWE Doppler Wind Experiment (Huygens science instrument)

E

E-ring Outermost Saturn ring

EGA Engine gimbal actuator

EGE Engine gimbal electronics

ESA European Space Agency

F

F-ring Narrow ring outward of A-ring

G

G-ring Diffuse ring at inner edge of E-ring

GCMS Gas Chromatograph and Mass Spectrometer (Huygens science instrument)

GHz Gigahertz (billions of cycles per second)

H

H^+ Hydrogen atom with one electron removed, i.e., a proton

H_2^+ Hydrogen molecule with one electron removed

APPENDIX B

H_2O^+ Water molecule with one electron removed

HASI Huygens Atmospheric Structure Instrument (Huygens science instrument)

HC_3N Cyanoacetylene

HCN Hydrogen cyanide

He^{++} Helium atom with two electrons removed, i.e., an alpha particle

HGA High-gain antenna

I

INMS Ion and Neutral Mass Spectrometer (Orbiter science instrument)

IRI Inertial reference interface

IRU Inertial reference unit

ISS Imaging Science Subsystem (Orbiter science instrument)

J

JPL Jet Propulsion Laboratory

L

LGA1 Low-gain antenna #1

LGA2 Low-gain antenna #2

M

MAG Dual Technique Magnetometer (Orbiter science instrument)

MAPS Magnetospheric and Plasma Science

MIMI Magnetospheric Imaging Instrument

MP Magnetopause (outer edge of Saturn's magnetic field)

N

N^+ Nitrogen atom with one electron removed

N_2 Nitrogen molecule

NASA National Aeronautics and Space Administration

O

O^+ Oxygen atom with one electron removed

ODM Orbiter deflection maneuver

OH Hydroxyl radical

OH^+ Hydroxyl radical with one electron removed

ORS Optical remote sensing

P

PMS Propulsion module subsystem

PPS Power and pyrotechnics subsystem

R

RADAR Cassini Radar (Orbiter science instrument)

RFES Radio frequency electronics subsystem

RFS Radio frequency subsystem

RPWS Radio and Plasma Wave Science (Orbiter science instrument)

R_S Unit of distance in radii of Saturn (one R_S = 60,330 kilometers)

RSS Radio Science Instrument (Orbiter science instrument)

RTG Radioisotope thermoelectric generator

RWA Reaction wheel assembly

RWI Reaction wheel interface

APPENDIX B

S

S Bow shock sunward of Saturn's magnetic field

SED Saturn electrostatic discharge

SKR Saturn kilometric radiation

SOI Saturn orbit insertion

SRMU Solid Rocket Motor Upgrade (part of Titan IV launch vehicle)

SRU Stellar reference unit

SSE Sun sensor electronics

SSH Sun sensor head

SSP Surface Science Package (Huygens science instrument)

T

TCS Temperature control subsystem

U

UK United Kingdom

US United States

UVIS Ultraviolet Imaging Spectrograph (Orbiter science instrument)

V

VDE Valve drive electronics

VHSIC Very high-speed integrated circuit

VIMS Visible and Infrared Mapping Spectrometer (Orbiter science instrument)

VVEJGA Venus–Venus–Earth–Jupiter Gravity Assist

APPENDIX C

General Index

APPENDIX D

Solar System Characteristics

STATISTICS OF THE PLANETS

	Mercury	Venus	Earth	Mars	Jupiter	Saturn	Uranus	Neptune	Pluto
Mean Distance from Sun									
Astronomical Units	0.387	0.723	1.000	1.524	5.203	9.555	19.218	30.110	39.545
Millions of kilometers	57.9	108.2	149.6	227.9	778.3	1429.4	2871.0	4504.3	5913.5
Orbital Period, days	87.97	224.7	365.26	686.98	4332.7	10759.5	30685	60190	90800
Diameter, kilometers									
Equatorial	4878	12104	12756	6794	142984	120660	51118	49528	2300
Polar	4879	12104	12713	6760	133829	107629	49586	48044	2300?
Mass [Earth = 1]	0.055	0.815	1.000	0.107	317.92	95.184	14.54	17.15	0.0022?
Density, t/cubic m	5.42	5.25	5.52	3.94	1.33	0.69	1.27	1.64	2.03
Gravity [Earth = 1]	0.38	0.90	1.00	0.38	2.53	1.07	0.91	1.14	0.07
Rotation Period, days	58.646	243.01	23.9345	24.6229	9.9249	10.6562	17.24	16.11	6.3872
Orbit Inclination, degrees	7.0	3.4	0.0	1.8	1.3	2.5	0.8	1.8	17.1

Footnotes

1 Astronomical Unit [AU] = 149,597,870.694 kilometers

Data from the Observer's Handbook 1997, courtesy of The Royal Astronomical Society of Canada and the Cassini Physical Constants document.

MAGNETOSPHERES OF THE PLANETS

	Mercury	Earth	Jupiter	Saturn	Uranus	Neptune
Field Strength at Equator, gauss•R_p^3	0.0033	0.3076	4.28	0.218	0.228	0.133
Dipole Tilt to Rotational Axis, degrees	169	11.4	9.6	<1	58.6	46.8
Typical Distance to Sunward "Nose" of Magnetosphere, planet radii	~1	10	65	20	18	25

APPENDIX D

THE SATELLITES – A COMPARISON

Planet	No., Designation	Name	Diameter, kilometers	Distance from Planet, kilometers – planet radii	Discovery
Saturn	XVIII, 1981S13	Pan	20	133583 – 2.214	1990
	XV, 1980S28	Atlas	37.0×34.4×26.4	137640 – 2.281	1980
	XVI, 1980S27	Prometheus	148×100×68	139350 – 3.310	1980
	XVII, 1980S26	Pandora	110×88×62	141700 – 2.349	1980
	XI, 1980S3	Epimetheus	138×110×110	151422 – 2.510	1966
	X, 1980S1	Janus	194×190×154	151472 – 2.511	1966
	I	Mimas	418.2×392.4×382.8	185520 – 3.075	1789
	II	Enceladus	512.6×494.6×489.2	238020 – 3.945	1789
	III	Tethys	1071.2×1056.4×1051.6	294660 – 4.884	1684
	XIII	Telesto	30×25×15	294660 – 4.884	1980
	XIV	Calypso	30×16×16	294660 – 4.884	1980
	IV	Dione	1120	377400 – 6.256	1684
	XII, 1980S6	Helene	35	378400 – 6.266	1980
	V	Rhea	1528	527040 – 8.736	1672
	VI	Titan	5150	1221850 – 20.253	1655
	VII	Hyperion	360×280×225	1481100 – 24.550	1848
	VIII	Iapetus	1436	3561300 – 59.030	1671
	IX	Phoebe	230×220×210	12952000 – 214.7	1898
Earth	–	Moon	3476	384500 – 60.285	–
Mars	I	Phobos	21	9400 – 2.767	1877
	II	Deimos	12	23500 – 6.918	1877
Jupiter	I	Io	3630	422000 – 5.903	1610
	II	Europa	3140	671000 – 9.386	1610
	III	Ganymede	5260	1070000 – 14.967	1610
	IV	Callisto	4800	1885000 – 26.367	1610
Neptune	I	Triton	2700	5510000 – 222.491	1846

Data from the Observer's Handbook 1997, courtesy of The Royal Astronomical Society of Canada and the Cassini Physical Constants document.

APPENDIX D

RING SYSTEMS OF THE OUTER PLANETS				
Planet	Equatorial Radius kilometers – radii	Ring Designation	Ring Inner Edge kilometers – radii	Ring Outer Edge kilometers – radii
Jupiter	71490 – 1.00	Halo	71490 – 1.00	122130 – 1.71
		Main	122130 – 1.71	129130 – 1.81
		Gossamer	129130 – 1.81	210000 – 2.94
Saturn	60330 – 1.00	D	66970 – 1.11	74510 – 1.24
		C	74510 – 1.24	92000 – 1.53
		B	92000 – 1.53	117580 – 1.95
		A	122170 – 2.03	136780 – 2.27
		F	140180 – 2.32	(width 50 kilometers)
		G	170180 – 2.82	(width variable)
		E	181000 – 3	483000 – 8
Uranus	25560 – 1.00	1986U2R	37000 – 1.45	39500 – 1.55
		6	41837 – 1.64	(width 2 kilometers)
		5	42235 – 1.65	(width 2 kilometers)
		4	42571 – 1.67	(width 2 kilometers)
		Alpha	44718 – 1.75	(width 10 kilometers)
		Beta	45661 – 1.79	(width 10 kilometers)
		Eta	47176 – 1.85	(width 1 kilometers)
		Gamma	47626 – 1.86	(width 3 kilometers)
		Delta	48303 – 1.89	(width 6 kilometers)
		Lambda	50024 – 1.96	(width 2 kilometers)
		Epsilon	51149 – 2.00	(width 20–90 kilometers)
Neptune	24765 – 1.00	Galle	41900 – 1.69	(width 15 kilometers)
		Leverrier	53200 – 2.15	(width 30 kilometers)
		Lassell	53200 – 2.15	(narrow)
		Arago	53200 – 2.15	59000 – 2.38
		Unnamed	61950 – 2.50	(indistinct)
		Adams	62900 – 2.54	(width 50 kilometers; has 3 or more denser arcs)

APPENDIX E

Program & Science Management

PROGRAM MANAGEMENT

Name	Title	Name	Title
E. Huckins	Program Director	R. Brace	Product Assurance Manager
R. Spehalski	Program Manager	D. Kindt	Huygens Probe Integration Manager
R. Draper	Deputy Program Manager	W. Fawcett	Science Instruments Manager
H. Hassan	Huygens Project Manager	P. Doms	Mission and Science Operations Manager
C. Kohlhase	Science and Mission Design Manager	J. Gunn	Mission and Science Operations Development Manager
T. Gavin	Spacecraft System Manager	A. Tavormina	Mission and Science Operations Operations Manager
C. Jones	Spacecraft Development Manager	J. Leising	Program Engineering Manager
G. Parker	Spacecraft Integration and Test Manager	R. Wilcox	Launch Approval Engineering Manager

SCIENCE LEADERSHIP

Name	Title	Name	Title
H. Brinton	Program Scientist	D. Gautier	Interdisciplinary Scientist
D. Matson	Project Scientist	T. Gombosi	Interdisciplinary Scientist
J.-P. Lebreton	Huygens Project Scientist	J. Lunine	Interdisciplinary Scientist
L. Spilker	Deputy Project Scientist	T. Owen	Interdisciplinary Scientist
E. Miner	Science Manager	F. Raulin	Interdisciplinary Scientist
G. Israel	ACP Principal Investigator	L. Soderblom	Interdisciplinary Scientist
D. Young	CAPS Principal Investigator	D. Strobel	Interdisciplinary Scientist
E. Grün	CDA Principal Investigator	C. Porco	ISS Team Leader
V. Kunde	CIRS Principal Investigator	D. Southwood	MAG Principal Investigator
M. Tomasko	DISR Principal Investigator	S. Krimigis	MIMI Principal Investigator
M. Bird	DWE Principal Investigator	C. Elachi	RADAR Team Leader
H. Niemann	GCMS Principal Investigator	D. Gurnett	RPWS Principal Investigator
M. Fulchignoni	HASI Principal Investigator	A. Kliore	RSS Team Leader
H. Waite	INMS Team Leader	J. Zarnecki	SSP Principal Investigator
M. Blanc	Interdisciplinary Scientist	L. Esposito	UVIS Team Leader
J. Cuzzi	Interdisciplinary Scientist	R. Brown	VIMS Team Leader

APPENDIX F

Bibliography

A

Alexander, A. F O'D. The Planet Saturn: A History of Observation, Theory and Discovery. MacMillan. 1962.

Atkinson, D. H., J. B. Pollack and A. Seiff. Measurement of a Zonal Wind Profile on Titan by Doppler Tracking of the Cassini Entry Probe. Radio Science, Vol 25. Sep–Oct 1990.

Atreya, S. Atmospheres and Ionospheres of the Outer Planets and Their Satellites. Springer Verlag. 1986.

Atreya, S. K., J. H. Waite, Jr, T. M. Donahue, A. F Nagy and J. C. McConnell. Theory, Measurements, and Models of the Upper Atmosphere and Ionosphere of Saturn. Saturn (T. Gehrels and M. S. Matthews, Eds). University of Arizona Press, Tucson, Arizona. 1988.

B

Bagenal, F Giant planet magnetospheres, Ann. Rev. Earth Planet. Sci. 20. 1992.

Barnard, E. E. Additional Observations of the Disappearance and Reappearances of the Rings of Saturn in 1907–08 Made with the 40-in Refractor of the Yerkes Observatory. MNRAS 68. 1908.

Barnard, E. E. Eclipse of Iapetus, the VIII Satellite of Saturn on Nov 1, 1889. Pubs of the Astronomical Society of the Pacific 1. 1889.

Barnard, E. E. Micrometrical Measures of the Ball and Ring System of the Planet Saturn and Measures of the Diameter of his Satellite Titan Made with the 36-inch Equatorial of the Lick Observatory. MNRAS 55. 1895.

Barnard, E. E. Observations of Saturn's Ring at the Time of Disappearance in 1907 Made with the 40-in Refractor of the Yerkes Observatory. MNRAS 68. 1908.

Barnard, E. E. Observations of the Eclipse of Iapetus in the Shadows of the Globe, Crape Ring and Bright Ring of Saturn, Nov 1, 1889. MNRAS 50. 1890.

Barnard, E. E. Transparency of the Crape Ring of Saturn and other peculiarities as shown by the observations of the eclipse of Iapetus on Nov 1, 1889. Astronomy and Astro-Physics 1. 1892.

Beatty, J. K., B. O'Leary and A. Chaikin (Eds). The New Solar System, 3rd ed. Sky Publishing Corp., Cambridge, Massachusetts. 1990.

Bertotti, B. An Introduction to the Cassini–Huygens Mission, Nuovo Cimento C, Serie 1, Vol 15C, No 6. Nov–Dec 1992.

Borderies, N., P Goldreich and S. Tremaine. Unsolved Problems in Planetary Rings. Planetary Rings (R. Greenberg and A. Brahic, Eds). University of Arizona Press, Tucson, Arizona. 1984.

Bosh, A.S. and A. S. Rivkin. Observations of Saturn's Inner Satellites During the May 1995 Ring-Plane Crossing. Science 272. 1996.

Burns, J. A., M. R. Showalter and G. E. Morfill The Ethereal Saturn's Rings of Jupiter and Saturn. Planetary Rings (R. Greenberg and A. Brahic, Eds). University of Arizona Press, Tucson, Arizona. 1984.

Burns, J. and M. Matthews (Eds). Satellites. University of Arizona Press, Tucson, Arizona. 1986.

C

Chaisson, E. and S. McMillan. Astronomy Today, 2nd ed. Prentice Hall, New Jersey. 1996.

Coates, A. J., C. Alsop, A. J. Coker, D. R. Linder, A. D. Johnstone, R. D. Woodliffe, M. Grande, A. Preece, S. Burge and D. S. Hall. The Electron Spectrometer for the Cassini Spacecraft. J. British Interplanetary Society, Vol 45, No 9. Sep 1992.

Comoretto, G., B. Bertotti, L. Iess and R. Ambrosini. Doppler Experiments with Cassini Radio System. Nuovo Cimento, Vol 15C, No 6. Nov–Dec 1992.

Connerney, J. E. P., L. Davis Jr and D. L. Chenette. Magnetic field models. Saturn (T. Gehrels and M. S. Matthews, Eds). University of Arizona Press, Tucson, Arizona. 1988.

Cuzzi, J. N. Evolution of Planetary Ring-moon Systems. Earth, Moon and Planets 67. 1995.

Cuzzi, J. N., et al. Saturn's Rings: Properties and Processes. Planetary Rings (R. Greenberg and A. Brahic, Eds). University of Arizona Press, Tucson, Arizona. 1984.

D

Davis, L. and E. J. Smith. A model of Saturn's Magnetic Field Based on All Available Data. J. Geophys. Res. 95. 1990.

Dermott, S. F. Dynamics of Narrow Rings. Planetary Rings (R. Greenberg and A. Brahic, Eds). University of Arizona Press, Tucson, Arizona. 1984, pp 589–637.

Dones, L. The Rings of the Outer Planets (preprint). 1997.

E

Ehrenfreund, P., J. J. Boon, J. Commandeur, C. Sagan, W. R. Thompson and B. Khare. Analytical Pyrolysis Experiment of Titan Aerosol and Analogues in Preparation for the Cassini Huygens Mission. COSPAR Plenary Meeting, Aug 28–Sep 5, 1992, and Advances in Space Research, Vol 15, No 3. Mar 1995.

Elachi, C., E. Im, L. E. Roth and C. L. Werner. Cassini Titan Radar Mapper. IEEE Proc, Vol 79. Jun 1991, pp 867–880. Esposito, L. W. Understanding Planetary Rings. Annual Review of Earth and Planetary Science 21. 1993.

Esposito, L. W., J. N. Cuzzi, J. B. Holberg, E. A. Marouf, G. L. Tyler and C. C. Porco. Saturn's Rings: Structure, Dynamics, and Particle Properties. Saturn (T. Gehrels and M. S. Mathews, Eds). University of Arizona Press, Tucson, Arizona. 1988.

European Space Agency. Announcement of Opportunity, Cassini Mission: Huygens. ESA SCI(89)2. Oct 1989.

European Space Agency. Cassini Saturn Orbiter and Titan Probe, ESA/NASA Assessment Study, ESA REF: SCI(85)1. Aug 1985.

European Space Agency. Cassini Saturn Orbiter and Titan Probe, Report on the Phase A Study, ESA SCI(88)5. Oct 1988.

European Space Agency. The Atmospheres of Saturn and Titan. Proceedings International Workshop, Alpbach, Austria, Sep 16–19, 1985. ESA SP-241. ESA, Paris, France. 1985.

F

Flanagan, S. and F. Peralta. Cassini 1997 VVEJGA Trajectory Launch/Arrival Space Analysis. AAS Paper 93-684. AAS/AIAA Astrodynamics Conference, Victoria, Canada. Aug 1993.

Formisano, F., A. Adriani and G. Bellucci. The VNIR–VIMS Experiment for CRAF/Cassini – Visual Near-IR Mapping Spectrometer. Nuovo Cimento C, Serie 1, Vol 15C, No 6. Nov–Dec 1992.

Franklin, F., M. Lecar and W. Wiesel. Ring Particle Dynamics in Resonances. Planetary Rings (R. Greenberg and A. Brahic, Eds). University of Arizona Press, Tucson, Arizona. 1984.

G

Galopeau, P. and P. Zarka. Evidence of Saturn's Magnetic Field Anomaly from Saturnian Kilometric Radiation High Frequency Limit. J. Geophys. Res. 96. 1991.

APPENDIX F

Galopeau, P H. M., P Zarka and D. L. Quéau Source location of Saturn's kilometric radiation: The Kelvin-Helmholtz instability hypothesis. J. Geophys. Res. 100. 1995.

Gehrels, T and M.S. Matthews (Eds). Saturn. University of Arizona Press, Tucson, Arizona. 1984.

Gibbs, R. Cassini Spacecraft Design. Cassini/Huygens: A Mission to the Saturnian System (L. Horn, Ed), Proc. SPIE 2803. 1996.

Goldreich, P and S. D. Tremaine. Towards a Theory for the Uranian Rings. Nature 277. 1979.

H

Harris, A. W The Origin and Evolution of Planetary Rings. Planetary Rings (R. Greenberg and A. Brahic, Eds). University of Arizona Press, Tucson, Arizona. 1984.

Hartmann, W K. Moons and Planets, 2nd ed. Wadsworth, Belmont, California. 1983.

Hassan, H., C. McCarthy and D. Wyn-Roberts. Huygens – A Technical and Programmatic Overview. ESA Bulletin No 7. Feb 1994.

Horn, L. J. (Ed). Cassini/Huygens: A Mission to the Saturnian System. Proc. SPIE 2803. 1996.

J

Jaffe, L. D. and J.-P Lebreton. The CRAF/Cassini Science Instruments, Spacecraft, and Missions. IAF Paper 90–436, Oct 1990, and Space Technology Vol 12. 1992.

Jaffe, L. D. and L. M. Herrell. Cassini/Huygens Science Instruments. Cassini/Huygens: A Mission to the Saturnian System (L. Horn, Ed), Proc. SPIE 2803. 1996.

Jones, C. P Cassini Program Update. AIAA Preprint 93–4745. 1993.

Juergens, D., J. Duval, Puget, A. Soufflot, V Formisano and V Tarnopolsky. A Common Design Imaging Spectrometer for Planetary Exploration. AIAA/JPL Conference on Solar System Exploration. 1989.

K

Kaiser, M. L., M. D. Desch, W.S. Kurth, A. Lecacheux, F Genova, B. M. Pedersen and D. R. Evans. Saturn as a Radio Source. Saturn, (T. Gehrels and M. S. Matthews, Eds). University of Arizona Press, Tucson, Arizona. 1988.

Keeler, J. E. First Observations of Saturn with the 36-inch Equatorial of the Lick Observatory, Sidereal Messenger 7. 1888.

Kohlhase, C. The Cassini Mission to Saturn – Meeting With a Majestic Giant. Planetary Report, Vol 13, No 4. Jul–Aug 1993.

Krimigis, S. M. A Post-Voyager View of Saturn's Environment. Johns Hopkins APL Technical Digest 3. 1982.

Kunde, V., G. Bjoraker, J. Brasunas, B. Conrath, F M. Flasar, D. Jennings, P Romani, R. Maichle, D. Gautier and M. Abbas. Infrared Spectroscopic Remote Sensing from the Cassini Orbiter. Optical spectroscopic instrumentation and techniques for the 1990s – Applications in astronomy, chemistry, and physics. Proc. Meeting, Las Cruces, New Mexico, Jun 4–6, 1990, SPIE. 1990.

Kurth, W. S. and D. A. Gurnett. Plasma Waves in Planetary Magnetospheres. J. Geophys. Res. 96. 1991.

L

Lebreton, J.-P and D. L. Matson An Overview of the Cassini Mission. Nuovo Cimento C, Vol. XV #6. 1992.

Lebreton, J.-P and D. L. Matson. An Overview of the Cassini Mission. Nuovo Cimento C, Serie 1, Vol 15C, No 6, Nov–Dec 1992.

Lebreton, J.-P Cassini, 1991, A Mission to Saturn and Titan. Proceedings of the 24th ESLAP Symposium, Formation of Stars and Planets, and the Evolution of the Solar System, Friedrichshafen, Sep 17–19, 1990. ESA SP-315.

Lebreton, J.-P M. Verdant and R. D. Willis. Huygens – The Science Payload and Mission Profile. ESA Bulletin No 7. Feb 1994.

Lunine, J. Does Titan Have Oceans? American Scientist, Vol 82, 1994.

Lyons, L. R. and D. J. Williams. Quantitative Aspects of Magnetospheric Physics. D. Reidel/Kluwer, Boston, Massachusetts. 1984.

M

McClintock, W E., G. M. Lawrence, R. A. Kohnet and L. W Esposito. Optical Design of the Ultraviolet Imaging Spectrograph for the Cassini Mission to Saturn. Instrumentation for Planetary and Terrestrial Atmospheric Remote Sensing. Proc. Meeting, San Diego, California, Jul 23–24, 1992, SPIE. 1992.

Mendis, D. A., J. R. Hill, W-H. Ip, C. K. Goertz and E. Grün, 1988. Electrodynamic Processes in the Ring System of Saturn. Saturn (T. Gehrels and M. S. Matthews, Eds). University of Arizona Press, Tucson, Arizona. 1988.

Mitchell, D. G., A. F Cheng, S. M. Krimigis, E. P Keath, S. E. Jaskulek, B. H. Mauk, R. W McEntire, E. C. Roelof, D. G. Williams, K. C. Hsieh and V. A. Drake. INCA: the ion neutral camera for energetic neutral atom imaging of the Saturnian magnetosphere. Optical Engineering 32. 1993.

Moore, P and G. Hunt. Atlas of the Solar System. Rand McNally, New York. 1984.

Morrison, D. Voyages to Saturn. National Aeronautics and Space Administration, NASA SP-451. Superintendent of Documents, U.S. Government Printing Office, Washington, D.C. 1982.

Murray, C. D. Planetary Ring Dynamics, Phil. Transactions of the Royal Society. London A 349. 1994.

Murray, C. D. The Cassini Imaging Science Experiment. J. British Interplanetary Society, Vol 45, No 9. Sep 1992.

National Aeronautics and Space Administration. Announcement of Opportunity, Cassini Mission: Saturn Orbiter. A.O. No. OSSA-1-89, Oct 10, 1989.

National Aeronautics and Space Administration. Announcement of Opportunity, Facility Instrument Science Team for the Ion and Neutral Mass Spectrometer on the Cassini Saturn Orbiter. A.O. No. OSSA-1-91, May 20, 1991.

N

Nirchio, F, B. Pernice, L. Borgarelli, C. Dionisio and R. Mizzoni. Cassini Radar Radio-Frequency Subsystem Design Description and Performance Analysis. 12th Annual International Geoscience and Remote Sensing Symposium, May 26–29, 1992, IEEE. 1992, Vol 1.

O

Osterbrock, D. E. and D. P Cruikshank. J.E. Keeler's Discovery of a Gap in the Outer Part of the A Ring. Icarus 53. 1983.

Osterbrock, D. E. Keeler's Gap. Science 209. 1980.

P

Peralta, F and J. Smith. Cassini Trajectory Design Description. AAS Paper 93-568. AAS/AIAA Astrodynamics Conference, Victoria, Canada. Aug 1993.

APPENDIX F

Peralta, F and S. Flanagan. Cassini Interplanetary Trajectory Design. Control Engineering Practice. Vol 3, No 11. 1995.

Porco, C. C. and G. E. Danielson. The Periodic Variations of Spokes in Saturn's Rings. Astronomical J. 87. 1982. 826–833. Porco, C. C., 1995. Highlights in Planetary Rings. Reviews of Geophysics, Supplement. 1995.

R

Ratcliffe, P R., J. A. M. McDonnell, J. G. Firth and E. Grün. The Cosmic Dust Analyser. J. British Interplanetary Society, Vol 43. 1992.

Reilly, T., K. Klaasen and S. Collin. Imaging System for MMII Spacecraft. AIAA/JPL Conference on Solar System Exploration. 1989.

Richardson, J. D. Thermal Ions at Saturn: Plasma Parameters and Implications. J. Geophys. Res. 91. 1986.

Rothery, D.A. Satellites of the Outer Planets. Clarendon Press, Oxford, U.K. 1992.

S

Scarf, F L., L. A. Fank, D. A. Gurnett, L. J. Lanzerotti, A. Lazarus and E. C. Sittler, Jr Measurements of plasma, plasma waves and suprathermal charged particles in Saturn's inner magnetosphere. Saturn, (T. Gehrels and M. S. Matthews, Eds). University of Arizona Press, Tucson, Arizona. 1988.

Schardt, A. W, K. W Behannon, R. P Lepping, J. FCarbary, A. Eviatar and G. L. Siscoe, 1988. The outer magnetosphere. Saturn, (T. Gehrels and M. S. Matthews, Eds). University of Arizona Press, Tucson, Arizona. 1988.

Science. Volume 212. Apr 10, 1981.

Science. Volume 215. Jan 29, 1982.

Sheehan, W The Immortal Fire Within: The Life and Work of Edward Emerson Barnard. Cambridge University Press. 1995.

Shu, F H. Waves in Planetary Rings. Planetary Rings (R. Greenberg and A. Brahic, Eds). University of Arizona Press, Tucson, Arizona. 1984.

Sittler, E. C., Jr, K. W Ogilvie and J. D. Scudder Survey of Low-Energy Plasma Electrons in Saturn's Magnetosphere: Voyagers 1 and 2. J. Geophys. Res. 88. 1983.

Smith, E. J., L. Davis Jr, D. E. Jones, P J. Coleman, D. S. Colburn, P. Dyal and C. P. Sonnett. Saturn's magnetosphere and its interaction with the solar wind. J. Geophys. Res. 85. 1980.

Southwood, D. J., A. Balogh and E. J. Smith. Dual Technique Magnetometer Experiment for the Cassini Orbiter Spacecraft. J. British Interplanetary Society, Vol 45, No 9. Sep 1992.

Spehalski, R. J. Cassini Mission to Saturn. Cassini/Huygens: A Mission to the Saturnian System (L. Horn, Ed), Proc. SPIE 2803. 1996.

Stone, E. C. and TC. Owen. The Saturn System. Saturn, (T. Gehrels and M. S. Matthews, Eds). University of Arizona Press, Tucson, Arizona. 1988.

Stone, E. C. Future Studies of Planetary Rings by Space Probes. Planetary Rings (R. Greenberg and A. Brahic, Eds). University of Arizona Press, Tucson, Arizona. 1984.

V

van Helden, A. Rings in Astronomy and Cosmology, 1600–1900. Planetary Rings (R. Greenberg and A. Brahic, Eds). University of Arizona Press, Tucson, Arizona. 1984.

van Helden, A. Saturn through the Telescope: A Brief Historical Survey. Saturn (T. Gehrels and M. S. Mathews, Eds). University of Arizona Press, Tucson, Arizona. 1984.

W

Walch, M., D. Juergens, S. Anthony, Nicholson and D. Phillip Stellar Occultation Experiment with the Cassini VIMS Instrument. SPIE. 1992.

Wolf, A. and J. Smith. Design of the Cassini Tour Trajectory in the Saturnian System. Control Engineering Practice, Vol 3, No 11. 1995.

Wolf, A. Touring the Saturnian System. Cassini/Huygens: A Mission to the Saturnian Systems (L. Horn, Ed). Proceedings of the SPIE 2803. 1996.

Woolliscroft, L. J. C., W.M. Farrell, H. St. C. Alleyne, D. A. Gurnett, D. L. Kirchner, W. S. Kurth and J. A. Thompson. Cassini Radio and Plasma Wave Investigation: Data Compression and Scientific Applications. J. British Interplanetary Society, Vol 46, No 3. Mar 1993.

www.ingramcontent.com/pod-product-compliance
Lightning Source LLC
Chambersburg PA
CBHW081724170526
45167CB00009B/3692